Unreal Objects

Digital Barricades:
Interventions in Digital Culture and Politics

Series Editors:

Professor Jodi Dean, Hobart and William Smith Colleges
Dr Joss Hands, Anglia Ruskin University
Professor Tim Jordan, University of Sussex

Also available:

Cyber-Proletariat:
Global Labour in the Digital Vortex
Nick Dyer-Witheford

Information Politics:
Liberation and Exploitation in the Digital Society
Tim Jordan

Unreal Objects

Digital Materialities, Technoscientific
Projects and Political Realities

Kate O'Riordan

PlutoPress
www.plutobooks.com

First published 2017 by Pluto Press
345 Archway Road, London N6 5AA

www.plutobooks.com

British Library Cataloguing in Publication Data
A catalogue record for this book is available from the British Library

ISBN   978 0 7453 3678 7   Hardback
ISBN   978 0 7453 3674 9   Paperback
ISBN   978 1 7868 0056 5   PDF eBook
ISBN   978 1 7868 0058 9   Kindle eBook
ISBN   978 1 7868 0057 2   EPUB eBook

Typeset by Stanford DTP Services, Northampton, England

Simultaneously printed in the United Kingdom and United States of America

Contents

List of Figures

Acknowledgements

This book partly comes out of my experiences of a technology assessment project called EPINET, conducted between 2012 and 2015. That research received funding from the European Community's Seventh Framework Programme (FP7/2007–2013) under grant agreement number 288971 (EPINET). Many of the people involved in the research consortium have had a formative influence on this book. In some cases where I've published on particular case studies with them, they enter into the writing. Specific acknowledgment and thanks goes to Dr Aristea Fotopoulou, with whom I co-authored work on biosensors, and some of her writing has undoubtedly made it on to the page in Chapter 3. Also acknowledgement and thanks go to Dr Neil Stephens, who conducted ethnographic work with the in vitro meat consortium and whose thinking and writing influences Chapter 5. Collaborative writing with Aristea and Neil was an aspect of EPINET that I really appreciated. Roger Strand and Kjetil Rommetveit led the EPINET consortium and their generosity and support was very significant in making the experience so interesting. I also learnt much from other members of the group and note my acknowledgments and thanks for the opportunity to work with them all. The full details of the project can be found at http://epinet.no.

During the same period I've been in an extended conversation with Professor Caroline Bassett and Professor Sarah Kember, along with whom my thinking about unreal objects has developed. They have commented on and encouraged earlier iterations of unreal objects, and presentations and writing around this topic. I also have to thank the Canadian Communication Association for inviting me to be their keynote speaker in 2015 during the development of this book. Having the time to give a longer talk was a luxury, and being pushed to make it work for an international audience helped me to bring these materials together. The feedback and discussions at the CCA conference were supportive and critical and helped me in my thinking. I really appreciated the opportunity to share this work in that community.

The book is also the product of earlier projects with Professor Maureen McNeil, Dr Joan Haran and Professor Jenny Kitzinger. It is a

viii · UNREAL OBJECTS

while since we worked on human cloning together, but the framework of thinking about how science is made in the media was formed in that work. Maureen and Joan have continued to encourage and support me since.

Dr Joan Haran has been a writing partner and source of support and friendship for the last decade, and she read and responded to multiple drafts of this book. Without her input I would not have got this to the point of publication and I have been continually inspired by her scholarship and thinking about feminism and science fiction as well as her friendship.

The book manuscript was completed during a period of leave from the University of Sussex and supported by the editors of this book series. Particular thanks go to Professor Tim Jordan who both authorized the leave and is one of the editors.

Also at Sussex, members of the Sussex Humanities Lab (directed by Professor Caroline Bassett) and affiliates, including Professor Sally Jane Norman, Kate Braybrooke, Stephen Fortune, Irene Fubara-Manuel, and participants in the Sussex-UCSC digital exchange, including Emile Devereaux, David Harris, Mary Agnes Krell, Gene Felice II and Professors Jennifer Parker and Sharon Daniel are all due acknowledgement and thanks for helping me think about biosensors and art and science, as well as being great people to work and play with. Stephen Fortune's work on the Quantified Self has also been influential in my thinking, especially about big data metaphors and the question of agency and driving when it comes to data.

Professor Jenny Reardon at UCSC has been influential in shaping some of my thinking through discussions about this material, especially genomics, but also about storytelling and meaning making. Her support over the years and recent inspirations from bandit biking in the last months of writing this draft have also helped me immeasurably.

Finally, much thanks and appreciation go to my family and friends who have put up with book angst for so long.

Series Preface

Crisis and conflict open up opportunities for liberation. In the early twenty-first century, these moments are marked by struggles enacted over and across the boundaries of the virtual, the digital, the actual and the real. Digital cultures and politics connect people even as they simultaneously place them under surveillance and allow their lives to be mined for advertising. This series aims to intervene in such cultural and political conjunctures. It will feature critical explorations of the new terrains and practices of resistance, producing critical and informed explorations of the possibilities for revolt and liberation.

Emerging research on digital cultures and politics investigates the effects of the widespread digitization of increasing numbers of cultural objects, the new channels of communication swirling around us and the changing means of producing, remixing and distributing digital objects. This research tends to oscillate between agendas of hope, that make remarkable claims for increased participation, and agendas of fear, that assume expanded repression and commodification. To avoid the opposites of hope and fear, the books in this series aggregate around the idea of the barricade. As sources of enclosure as well as defences for liberated space, barricades are erected where struggles are fierce and the stakes are high. They are necessarily partisan divides, different politicizations and deployments of a common surface. In this sense, new media objects, their networked circuits and settings, as well as their material, informational and biological carriers all act as digital barricades.

Jodi Dean, Joss Hands and Tim Jordan

Preface

1

Introduction: Problems With Objects

'Unreal Objects' as a title might seem like a contradiction. That is the point. This is a book about contradictory and competing realities. The world is full of technological objects that are naturalized and taken as a given. Accepting these objects in their own terms means that responding reactively to them is one of the only positions available. Objects orientate people, knowledge and worlds. The point of the book, then, is to disorientate some of these objects and look at ways of taking them in different terms. There is an imperative to look at the world and its phenomena in terms of objects, and to disavow other ways of knowing by prioritizing some objects over others. This appears in particular kinds of materialist thinking such as object orientated philosophy and acceler-ationism (Bogost 2006, 2012; Morton 2013; Williams and Srnicek 2013). I'm going to refer to this as object materialism. Materialism itself is not the issue at stake here – multiple kinds of material thinking contribute to knowing and intervening in the world. Feminist materialism, historical materialism, science studies and ecological materialisms are also influential in taking things seriously as both material and representa-tional. However, a particular kind of insistence on the object, in both the claims of technoscience and directions in academic and political thinking, are part of a problem to be addressed here. The way that object materialisms in the world of theory seem to mirror the claims of techno-science is striking; both insist on taking particular technological objects as a given, in their own terms. The book works to bring back a sense of objects as things in the making, mediated, unstable, not quite given, constantly deferred, and as part of the problem of always positing science and technology as the answer.

The book undoes this imperative to be object orientated by looking at what I'm referring to as unreal objects. Taking digital-media-materi-ality together amounts to a proposition that media objects mediate and make worlds, and that what counts as media and as material are political questions. Approaching emerging technoscientific projects as unreal objects is a way of challenging the imperative to find technological fixes for social issues, and the demand for everything to be an object.

The imperative to look at the world in terms of the agency of real objects operates across political discourse, technological, scientific and engineering fields, and philosophical and critical theory. For example, the US president Donald Trump promises to build a wall, delivering an agential object as a political solution. In the UK, Trident is offered as an object, a self-defined thing: it is what it is. Walls and weapons are offered as real political objects making cuts in the world. At the same time, in the register of philosophical and critical theory, leading thinkers tell us that there is a world of objects that appears to us directly, unmediated, and that we have to deal with this world reactively, in material terms.

The current focus given to objects and the idea that we can only deal with the reality of the world as it is given to us might be an abdication in bad faith. It leaves reaction as the only option and impels acceptance of multiple factors as just realities we have to deal with. However, realities are made up too, and the full capacity of 'made up' to mean manufactured, created, invented is important here. Objects are not just givens to which reaction is the only orientation. Politics are involved in the making of objects, realities and worlds. It seems to me that there are two types of object that are given to us as real: those that are construed as arising from the world, like bodies and mountains; and those that are made in the world, like iPhones and computers. Even though the latter are more obviously made up, manufactured, real things, they too are taken as inevitable. Their inevitability, high status and economic value mean that they outweigh other kinds of realities in a hierarchy of unreal objects. The status of technoscientific objects has a special role in securing the real: they are both made up and promise to remake other realities. Genomes will remake bodies, biosensors will remake homes and cities, smart grids will remake climates.

This is then a book about emerging technologies, new things that promise to remake other realities. Some of the examples are more emerging than others. Some haven't made it off the prospectus and others have already become part of everyday life. All the examples in this book can be thought of as big emerging technosciences, and the idea that they will all be realized in the world is a naturalized and deterministic story that I seek to disrupt. All are emerging in a moment in which the role of the media is central to the research into, and the development and delivery of, new technoscientific realities. The role of the media is folded into these projects in multiple ways. On the one hand, the role of public relations and creative media agencies is pervasive in the development of these projects from their very early stages. On the other hand, technoscientific objects themselves constitute processes of mediation, stabilizing temporary realities through media texts, devices, sequences and platforms. All of the examples join up technologies and bodies to create sites at which biological materials and informational technologies circulate, flow and mediate each other.

I use the term unreal here to try to emphasize hierarchies of reality and of materiality and to demonstrate differential materialities and realities. The unreal objects of the title are media materialities, objects which are given as real but also operate on a spectrum that includes what can also be thought of as immaterial, symbolic, insubstantial and unreal. Unreal objects are both a proposition and an approach: a proposition that objects that appear real are also made up; and an approach to emerging technologies that takes them as objects and discourses, material and symbolic, imaginary and actual. They are contradictory things in the world that can serve as reminders of the contradictions of given realities. This is to point to forms of intervention, thereby disrupting the narrative of the inevitable world given to us in which we can only react.

The premise of this book is that political legitimacy is negotiated through science and technology taken as objects, that mediation is central in materializing this authority as real, but that other stories can be told which undo the objects of technoscience as they are given. Emerging technologies become nodes of contestation about what collective investments should be made and what common futures are desirable, and as such they are political objects. However, the question of which objects come to accumulate that political gravity, or to assume a reality, has as much to do with the media life of these objects as anything else.

SOME BACK STORY: WORKING WITH EMERGING TECHNOLOGIES

I've been thinking about unreal objects for some time, and some specific experiences will help to tell a story about how this developed into a proposition and an approach. The first is an anecdote about a dinner conversation. I was working on a three-year project about the economic and social aspects of genomics. This was just after the Human Genome Project had been completed and some two decades into the emergence of genomics as a global big science endeavour. At the annual project conference dinner one of my colleagues observed jokingly to a genome scientist that she wasn't sure if she believed in the genome. After all, she noted, you can't see it or show it to me.

We had both recently read the novel *Life* by Gwyneth Jones, in which the protagonist conjures a strand of DNA from onions and washing-up liquid. DNA you can touch and see. Genes and genomics on the other hand are not apparent to the eye. Genomes are only manifest as objects as sequences of data, three billion base pairs per genome. You can look but you can't really touch. The human genome is printed out as a sequence of letters in a book in the Wellcome library and you can touch the book – but this is a book, a media object, not a genome. The genome sequence is likewise a sequence not a genome. On one level it is hard to believe in genomes, and this story about scepticism expressed at the centre of genomic research is refreshing. On another level, a huge amount of attention, investment, work and media production has gone into making genomes objects. This realization and materialization has been complex, produced through networks of objects, actors and processes of mediation over many decades. They have real effects on people's lives, from the careers of scientists, to the experiences of research subjects and patients.

At the time of this conversation my attachment to genomes was abstract. I'd been working on the economic and social aspects of genomics as a media analyst for some time and continued to do so for a decade. Towards the end of that time my attachment became more passionate when I discovered that my mother, my sister and I had a relatively rare genetic condition. Whether passionately or abstractedly invested, it is clear that genomes occupy such an important position that world leaders have claimed they are the language of god, and billions of pounds, dollars and other currencies have been poured into them. Although, as other scientific fields come into (re)ascendance in the early twenty-first century (physics and neuroscience in particular), it is also clear that perhaps there are fashions in the sciences as elsewhere. The £11 billion spent on the Hadron Collider, which opened in 2008, overshadowed the estimated spend of £5 billion on the Human Genome Project completed in 2000, or thereabouts.[1] Science and their technologies rise to prominence, rule the day and move on. However, as mediations they don't disappear, they reanimate and remediate (Bolter and Grusin

---

1 The draft genome was announced in 2000 but the Human Genome Project wasn't officially competed until 2004.

1998). For example, neuroscience reanimates psychology, and genomics remediated questions about the effect of nuclear and chemical warfare on populations (Higuchi 2010; Cook-Deegan 1991).

Another key experience that shaped this book was my involvement in the technology assessment project EPINET.[2] As part of a larger consortium, I led the media analysis strand of the project alongside people working on environmental, economic, legal, socio-technical and ethical aspects. At the time I was surprised that the research objects in each strand of the project were media materials. The basic units of analysis were texts produced about the technologies. Where there were prototypes, trials or pilots they were communicated through reports, images, texts, conferences, conversations, as well as assemblages of actors, relations and objects. We had been commissioned to look at technological objects, which although designated as emerging, were defined as things in the world. The emerging technologies were already given to us as objects, in relation to which assessment was reactive.

In this project the media analysis was distinct because we were looking at public and audience engagement and mediated visions and imaginaries. However, our strongest contribution was in some ways the reminder that other forms of assessment were also looking at visions. We compared use and take up with prospective visions, and focused on questions about the forms of media production and consumption involved. However, the objects kept shifting, and my overriding impression coming out of that project was that these emerging technologies, which included in vitro meat, biosensors and smart grids, were, above all, media objects. Things, and discourses, formations, tropes, figures, visions made up through media forms, and the attempts to define these as objects, were communicative, world-making processes that embedded the beliefs of those making, attending and investing in them.

The idea that technology is the materialization of cultural beliefs or is a cultural form is not a novel observation; it has been influential in both media and science and technology studies (Williams 1974; Latour 1991). That imaginaries are world-making is a proposition that has been examined in feminist approaches to technoscience, and especially in the work of Donna Haraway (1988, 1992, 1997). The proposition that we can only react to objects is at odds with these approaches to science and technology. Objects after all are orientating devices (Ahmed 2006), and

---

2 The epistemic networks project: epinet.no

to suggest that the objects of technoscience are unreal is to provide some disorientation as an intervention.

OBJECTS IN THE BOOK

The examples of unreal objects that are analysed in this book are: human genomics, biosensors, smart grids, in vitro meat, and de-extinction. The chapters are arranged around each of the listed examples, with in vitro meat and de-extinction considered in the same chapter. In the second chapter I focus on the case of Genomics England to discuss human genomics. Human genomics is a massive terrain and multiple books have been written about its economic, cultural and social aspects over the last two decades. In the spectrum of unreal objects considered here it is well established. Genomes are media objects which have a very high media presence and a digital media ontology. This is because genomes take the form of sequences, anchored in an imagined biological materiality to which there is a very strong ontological claim but no object. Genomes are digital media, or at least appear as such in sequence form, but as the chapter demonstrates, these sequences simultaneously appear and are deferred as objects, made relational through the imperative to collect them in large numbers. Human genomics brings human biology, genetics and informatics together. Chapter 2 explores some of the media work of Genomics England and sets it in the context of the political economy of sequencing. In doing so the chapter draws out the way genomes are made meaningful in this context, but also suggests that we need to think about them otherwise.

Each chapter looks at an example in terms of how it is given as an object and set up in a dominant or preferred form, but also looks at counter versions, alternatives and contradictions. In using this strategy I aim to bring an analysis of the objects together with the suggestion of alternative ways of understanding them. For example, Genomics England is an investment based on the promise of genomics to revolutionize biomedical health care; an alternative way of seeing this is to understand genomics as part of a digital economy, driving big data and sequence technology. It also offers investment in genetic editing technologies and the possibility of engineering species and it is important to bring this into focus when the question of NHS resources are at stake.

The third chapter is on biosensors. It looks more specifically at fitness tracking technologies, object devices that measure and quantify human movement, calorie consumption and sleep patterns. Biosensors, as a category, refer to a much wider range of technologies that sense and measure biological signals and create data streams based on these. They have application as scientific instruments, in climate science, health care and leisure. Examples include monitoring blood sugar for diabetes or measuring sweat for fitness training or chemical composition. An early example of an analogue biosensor is the so-called lie-detector or polygraph test which senses several biological signs including blood pressure, pulse, respiration and skin conductivity (Littlefield 2008). The rhythms of these signals were written out in patterns and subject to interpretation. In the examples explored here, these layers of collecting, recording and interpreting are condensed into a device, which provides a strong interpretative framework for the biological data collected. The chapter uses the example of fitness monitoring to look at how the mass-market roll out of such technologies has been taken up. It sets these objects alongside other forms of measuring and recording fitness in everyday life, by looking at diaries and letters in earlier periods. It also sets the market model of fitness tracking against digital art practices and alternative interventions into these technologies. Like the previous chapter, it does this to look both at the object as a mass-market product to which only a reactive response is offered, and at how it might be otherwise.

The fourth chapter on smart grids allows a different scale of unreal object to unfold. Smart grids are visions of alternative energy futures, which scale up to international networks. They are given as objects represented in diagrams and an industry. They are at the same time symbolic forms, extrapolating the network mode as a vision for energy. They have materialized as objects in the smart meter, which then stands in for the vision even as it embodies its contradiction. In smart-grid visions the existing national or local grids that distribute energy from one source to multiple consumers are transformed. The promised transformation is to a flexible grid with multiple energy sources, including renewables and consumer-produced forms of energy, in which smartness refers to computerized self-monitoring systems that use energy in optimal ways. To date smart grids are anchored in the roll out of smart meters, and the chapter examines how this is being conducted in the UK. In this roll out, attachment, love and nostalgia are engaged, and so the making of unreal objects as love objects is also explored. The love of technology and the enchantment of technological objects is central to unreal objects as a whole. I draw specifically on Bruno Latour's (1996) work on Aramis and ideas about the love of technology to illustrate this in relation to smart grids and to undo their abstractions.

The fifth chapter takes in vitro meat and de-extinction together. In vitro meat is a field in which tissue culturing is the basis for creating new meat forms, or meat outside of the animal. Tissue and cell cultures are grown in laboratory conditions in order to develop new meaty food products. De-extinction on the other hand is the cloning or genetic engineering of extinct (or nearly extinct) species in order to bring them back into being. In both cases the digitization of biological signals, and the dislocation of biological materials from embodied contexts to bio-technological ones, provides the basis for creating new bodies in the world. The objects discussed in this chapter are the temporary object of the in vitro meat burger, and the almost object of the de-extinct animal or cloned organism.

The three chapters on genomics, biosensors and smart grids are largely about digital inscription; that is, the making of things as digital forms: blood and tissue samples into genome sequences; biological functions into data; energy into computing infrastructure. The last two examples, in vitro meat and de-extinction, are about rendering digital materials into fleshly entities. All of the examples constitute a biodigital milieu because they involve multi-directional flows through biological and digital forms, but the direction of flow is perhaps more clearly biological to digital in the earlier chapters, and digital to biological in the later chapter.

MATERIAL AND IMMATERIAL: REAL AND UNREAL

In the last months of this project three phenomena came more clearly onto the horizon. Violent public attacks on black people, queers, Muslims, migrants and left-wing politicians came to the forefront of political and media attention in Europe and the United States; regressive political changes materialized further as the UK voted to leave the European Union and Donald Trump became president elect of the United States; and *Pokémon Go* emerged.

Pokémon Go demonstrates something of the enchantment of unreal objects. It is a game in which virtual Pokémon are hunted, captured and trained in augmented space. Players need a device (phone or tablet) with data. The on-screen view of Google Maps is populated with characters such that the screen appears to show a virtual world hidden from actual view, but in which actions in actual space create game-world effects. Walking through the city is augmented by the mobile screen to create a space where Pokémon appear and can be captured. The reach of *Pokémon Go* right now seems symptomatic of the attraction of unreal objects in a moment in which our capacity to care about real people's lives is uncertain. Pokémon is a much longer-term phenomenon and has seen mass popularity at other times (Allison 2003; Bainbridge 2013; Gibson 2002; Jordan 2004). However, its popularity at this moment has taken on a different, more diverse and ubiquitous form (Giddings 2016; Keogh 2016; Salen Tekinbas 2016). The current game utilizes augmented reality (AR). Largely confined to games, heritage, art and education projects since the 1990s, AR has found it difficult to establish broad market appeal. *Pokémon Go* has changed this entirely by introducing players to a user-friendly (although data-heavy) version of AR. Augmented reality offers another entanglement of mediated and real, remediating the actual environment as a game space in a layering that augments rather than separates out. This layering is similar to the 3D projection technology in the iMAX but AR is distributed across different spaces (locative or expanded media). It uses computational mobile devices as interfaces rather than the cinema.

The game is dependent on the idea that devices (phones, tablets), data and wi-fi are ubiquitous (Keogh 2016). It exploits and exacerbates a culture of acceptance around commodification, data mining and always-on ubiquitous devices. It is pleasurable, escapist and communal. It has come with its own scare stories about risks to players, and has garnered widespread media coverage. By July 2016, it had gained an estimated 30 million players. It's communal aspects seem like an antidote to the individualism of headsets or fitness devices. It provides pleasure and entertainment in a period marked by very dark political times. It also extends the colonizing force of the digital further, capturing more and more people in the intimate network of devices, data and media that constitute the contemporary commercial world. It directs our gaze and attention back to our devices, just as people were perhaps starting to look up from Facebook. It blends the actual and unreal, texturing the dreamscape of unreal objects further.

Pokémon Go is possibly easy to dismiss. It can be positioned as just a game, a fad, not serious, not real. It features animated characters which can be designated as low culture, mass culture, commodities, media animations, cartoons. However, the point of putting it alongside the more serious objects of technoscience is that they are all symptoms of the same layering of the digital and the biological. Unlike genomes, biosensors, smart grids and de-extinction, *Pokémon Go* has a sense of humour. It isn't shored up with promissory rhetoric about saving lives, worlds and futures but it crystalizes the same dynamics of being media all the way down, intervening in the real and having material effects. Alongside horrific political tensions, ongoing violence and nationalism, it looks as though the attractions of unreal objects are obvious. They offer romance, capture our attention and orient us towards alternative fantasies, futures and realties. The confluence of these things helps to illustrate one of the major issues of the moment: the question of how and what we care about. Black Lives Matter, the Orlando shootings, ongoing violent attacks on specific groups of people, and the disengagement with community-building ideals like the European Union all highlight the present constitutional crisis around care and attention for people and lives. This is a crisis of political constitution but also one about how the world is made up. While millions of people log onto *Pokémon Go* and love it, there is at the same time a lack of care for particular living bodies. In short, we have constructed systems of care for technoscientific objects, nurturing the growth of devices, platforms and data. At the same time as these objects seem to offer the possibility of coming together, we lack other structures of organization to come together and nurture people and their lives.

Pokémon Go is easy to locate as a media object, but this book is about bringing things that are less easily – or more uneasily – categorized as media objects to a media approach. It seems urgent that we recognize that things taken as objects are also media objects. We need to look beyond technoscientific enchantments to different realities and find ways to co-opt and divert these enchanting objects for less destructive projects.

MEDIA AND MATERIALISM

The use of 'unreal' in the title of this book is a provocation that gestures towards academic debates about materialism, objects and knowledge. In these debates there is an argument that too much attention has been given over to questions of meaning making or text and that what is urgent is the real, material world and particularly global warming (Williams and Srnicek 2013; Galloway et al. 2014; Bogost 2012; Morton 2013). However, the challenge that these interventions leave unresolved is that they assume that what is real is obvious and not already constructed through processes of meaning making. The sciences are offered up as offering a naturalized and direct form of knowledge, and technological objects as inevitable (Morton 2013).

Accelerationism and object oriented materialisms have also been contested (Cubonicks 2015; Asberg et al. 2015; Kember 2015). Cubonicks for example contests the offering of technological objects by acknowledging mediation: 'It is a world that swarms with technological mediation, interlacing our daily lives with abstraction, virtuality, and complexity' (Cubonicks, 2015: 0). There is also a productive nexus around feminist materialism, and feminist new materialisms have taken a different route to those of accelerated objects (Haraway 1997; Hinton and Van der Tuin 2014; V. Kirby, 1997, 2011; Asberg et al. 2015). These feminist materialisms share an attention to reworking textual-material dualities and to situation, identity and the way that 'the text, too, is a material reconfiguring' (Hinton and Van der Tuin, 2014). They take issue with the idea that we can just know what is real and act on it without attending to how particular versions of the real are known. However, at the same time there are strong voices that espouse getting real in ways that abdicate responsibility for that reality. The 'Accelerationist Manifesto' (Williams and Srnicek 2013) and versions of object orientated philosophy such as Timothy Morton's *Hyperobjects* (2013) are examples of this tendency to reduce things in the world to objects that can be known outside of mediation, and to which we can only react. These latter espousals are mirrored in the language of emerging technologies and technological innovation where the point is to put objects on the table and take the technological fix for granted.

I use the term unreal as a reminder that what counts as real is contingent, rather than common sense or shared, and this contingency matters. The aim is not to assert a real-unreal division, but rather to open up the categories of real and unreal and to demonstrate how things can be both real and unreal, material and immaterial, through overlapping dynamics. What is important is which things, experiences and meanings get to count as real at any given moment. The reality of things is contingent not just on the materials, practices and mediation involved; the question of whether something counts as an object or not is part of what is at stake. In other words, some objects are made to seem more real by those with interests in capturing attention and ontological affirmation; for example, Elon Musk claims that human space travel to Mars is an inevitable reality by taking the reusable space rocket as a given object. At the same time, some things are made more immaterial because they are politically inconvenient. For example, global warming is repeatedly stabilized and destabilized in political discourse. In another example, the refugee camps on the French/UK border have been unmade and remade through mediation as well as physical dismantlement. At one moment the place of refugees figured as vulnerable women and children, in the next as an encampment of underserving menacing men, and in the next dismantled entirely.

It is not enough to say we have to get real on the one hand or to dismiss things because they are media or texts on the other. Current conditions make it very difficult to be sure about which kinds of knowledge might be more robust, practical or meaningful than others. This comes at a time in which being able to account for the role of human action, inaction, invention and meaning in the world seems more urgent than ever. How people make decisions about the collective present and future is an ever more vexed question. The examples in this book raise questions about investment, attention and how collective imaginaries are created and enrolled. We can ask, for example, whether the NHS budget should be used to invest in genomics; whether humans should be in the business of synthesizing human genomes and cutting out genes; whether we should all monitor our health and activity, and if getting smart devices will improve our health chances; whether systems to create and manage big data will make lives more livable; whether we should create in vitro meat to ameliorate the impact of meat production, or if we should create new species; or whether smart meters and burying carbon will help ameliorate global warming. Emerging technosciences are harnessed to world making, or breaking, claims about better health, improved species conditions, global warming and world food distribution. In this mix, the question of whether claims, technologies or the conditions they address are real or not is important.

Some of the objects considered in this book appear to have fleshy materiality, for example in vitro meat, which has enough materiality to be eaten as food. Through the 'in vitro' it is dislocated from one set of materials and intertwined with another. That is to say, in vitro meat is dislocated from the bodies of animals and intertwined with those of the laboratory processes of tissue culturing. Some of these objects have media materiality, such as genomes, which are artefacts of digital media, existing as sequences and recordings of light refractions. However, this kind of media materiality is taken seriously as a given real. Genomes don't get trashed as media texts, and the work of interpreting genomes isn't designated as textual, or about meaning making. But it is. Genomes and *Pokémon Go* are not in the same discursive register when it comes to talking about them as things with social life, but they should be.

Genomes have a digital reality and materiality, which could be thought of as merely symbolic, insubstantial, immaterial. However, they are framed as having a special substantive reality and their materiality is distributed through multiple infrastructures. Genomes are never figured as representational media in their biomedical circulation. They are produced as realities. The accusation that to be of the media is to be of the symbolic, and therefore insubstantial or immaterial, is reserved for other media texts such as those of mass and popular culture. Genomes, although laden with textual metaphors (Nerlich and Hellsten 2004), are constructed as material things. Their immateriality in terms of their instantiation as media texts does not equate to immateriality in terms of importance. Perhaps we should not take this construction as seriously as we are told to. On the other hand, smart grids, which only exist in media forms as visions of energy futures and new distributions of electrical power and energy, are media texts. They are symbolic imaginaries, which inhere in media texts as a promise, in what might be referred to as the promissory rhetoric of public relations or advertisements for the energy industry.

Each of the examples in this book enables an unfolding of the object concerned in a different way. The term 'object' (like the use of 'unreal') is intended to be contentious. The anthropologist Tim Ingold (2012) argues that the designation 'object' creates a particular orientation towards things. He writes in a discussion of materials and matter:

> Anything we come across could, in principle, be regarded as either an object or a sample of material. To view it as an object is to take it for what it is: a complete and final form that confronts the viewer as a *fait accompli*. It is already made. Any further changes it may undergo, beyond the point of completion, consequently belong to the phase of use or consumption. (2012: 435)

This is the problem with how technological objects appear in the world. Even when they are emerging they appear as objects already here. Genomes, biosensors, smart grids and new biotechnological organisms are taken as real objects which can do certain things and demand responses. They also intensify and accelerate object making, particularly in the way that they turn things into data. Making, generating and collecting data is a way of making things objects. To frame these as unreal objects then is to bring them back to the life of things, processes, materials and meaning making. To perhaps re-do them as things.

Through this discussion the book addresses issues in the public understanding of science and technology. There are different structural tensions in the world of science and technology development on the one hand, and in critique and theorizing about science and technology on the other, but these tensions reflect each other. In the world of science and technology development there is a veneration of technological objects, and patterns of dismissal and antagonism towards media forms and their obvious made-up-ness. For example, journalists are often blamed for exaggeration, specious framings and inaccuracy in science reporting, even when the lead for such framings comes from science journals (Kitzinger 2006). On the other hand there has been a significant rise in the use of media work in science and technology, from science consultants in Hollywood film production (Kirby 2011) to the direct employment of creative agencies in early-stage research, and the use of media forms for engagement, promotion and dissemination. Thus, the increased deployment of media forms in tandem with an intensified veneration of objects occurs both in technological innovation and academic debate.

Media forms are used to incorporate technoscientific projects and to make them appear as objects in everyday life (O'Riordan 2010). This happens both at points where technologies are already consumer devices, and to encourage investment in technoscientific research at an early stage. It also happens when technologies are emerging, and even when they remain on the prospectus. These forms of media incorporation are disavowed at the same time that they are used as such. For example, one issue is that public relations, public engagement and advertising have become integral to science and technology development at a very early stage. However, there is a failure to acknowledge that this is part of the real business of science and to open up these relations to scrutiny and critical engagement. This intensified role of public relations is examined further in the chapters that follow.

Media technologies have always been bound up in the history of making knowledge in science (Cartwright 1995; Jordanova 1989), but the degree to which this has intensified makes it more complex and more difficult to think about how mediation and knowledge are made together. The examples in this book are both heavily and doubly mediated. Firstly, because of the production of multiple media texts in the emergence of these technologies, and secondly in the way that the technoscientific entities at the centre of each case also mediate. They are media forms communicating and instantiating knowledge, ideas and norms about health, identity, climate or species. In doing so they are part of the making of the conditions of possible knowledge and this makes them also deeply political.

This doubling of technoscientific mediation is also cut through with contested claims about which forms of science and technology are more real, or more promising than others. This is the basic question at the heart of governance in a technoscientific society. Which areas of research and development should be pursued? This question plays out in democratic, moral and economic directions: investors want to know if they will get an economic return, policy makers want to know if both economic and pragmatic benefits will come, people concerned with the idea of better worlds in moral and ethical terms want to know if good or bad things will result. Fear of hoaxes or hype are mixed up with hopes for technological fixes and control over the present and the future. Part of the problem is how to distinguish promising things from information about promise or pessimism, and lately the distinction between things and information itself seems to divide ways of thinking about this problem.

MEDIATION AND THE INFLATED IMPORTANCE OF MEDIA AND INFORMATION

The cultural context in many places has become more informational, data obsessed and mediatized. At the same time there has been a blurring of media forms, genres and structures such that public relations, advertising, promissory rhetoric and media framings of issues become taken as the things in themselves. As mediation has become more central, an anxiety with the real and a reliance on bodies, objects and materials as sources of truth has become more pronounced, but at the same time it is in the representation of objects and materials that their power to legitimate the real is made. For example, Andrew Barry argues in *Material Politics* (2013) that as things and materials become more important as guarantors of truth or reality they also become more informational, with multiple forms of documentation establishing the properties, definitions and interpretations of things.

The inflated importance of media and information can be read off the epic scale of digital media theory and the claims of new materialism. It can also be registered in the current emphasis on information, (big) data and the take up of media in everyday life. Unreal objects can be thought about along axes of media presence and ontological instability. Some objects have a high media presence and are ontologically unstable or have only media ontologies. Other objects have less media presence and high ontological stability, while others have a mixture of different kinds of media reality. As material politics become more central, the mediation of objects, their informatic capital, public relations and media forms become more important in deciding the reality of things. More and more things have media ontologies.

The claim to the epic in digital media theory can be seen in the influential titles in the field. For example, Mark Deuze's claim to *Media Life* (2012) is totalizing, and no less so is Jussi Parikka's *A Geology of Media* (2015), which promises to excavate millennia of media time. These claims to all of life, all of time, all of the horizon in digital media theory resonate with the claims of technoscience. From iPads to carbon capture, the promise of technological projects is to be relevant to all of life, all of the world. In this sense the public relations and promissory rhetoric of technoscience, and the grand ambitions of digital media theory, are both colonizing discourses that promise, or threaten, to take up the whole ground of everything. The inflated ego of digital media theory is part of the problem, and its reiteration on the same scale as the promissory rhetoric of technoscience indicates that it is caught up in the same frame.

There have been a number of attempts to rethink media studies in relation to pervasive mediation in the field. In *Excommunication: Three Inquiries in Media and Mediation*, Galloway, Thacker and Wark argue that just as media studies got interesting it became difficult to examine mediation: 'New kinds of limitations and biases have made it difficult for media scholars to take the ultimate step and study the basic conditions of mediation' (2014: 7). They go on to say that their aim is not to create a theory of mediation, 'for which there already exist a number of exemplars, but a theory of mediation as excommunication' (2014: 11). In this sense they aim to go beyond mediation and look both at its limits and insufficiency but also at what is beyond, and here they gesture towards the uncanny, 'xenocommunication' and death. There is an epic scale to this, and an abstraction or lack of situatedness that inflects this take on mediation.

Contrary to the claim above, that media scholars haven't looked at mediation, in fact processes of mediation are the subject of media studies, as Kember and Zylinska's *Life After New Media: Mediation as Vital Process* (2012) demonstrates. Kember and Zylinska make an original case for thinking about new media, and draw on a genealogy of writings that contribute to a consideration of mediation, including the traditions of media events and ritual, and Thacker's own work on biomedia (2004). The argument, explicit in the title, is that mediation is a vital process, by which they mean both vital as life (vitalism) and as essential. They also caution against taking objects in their own terms, arguing that the point is not the devices or gadgets of new media but the processes of mediation of which they are part.

In terms of everyday experience, it is hard to think of any aspect of society that does not pass through a media lens or is not made in a digital environment. However, as the editors of this Digital Barricades book series remind us, media are not evenly distributed phenomena: they offer enclosure and liberation unequally, engendering differing struggles with high stakes. This book takes up the figure of the unreal object in order to examine some of this uneven terrain and to introduce modes of differentiating kinds of media and kinds of material. On the one hand media are everywhere, on the other the scale of media requires thinking about the conditions of this, not just seeing it as additive.

DATA, DISTANCE AND SITUATED READING

Social commentary about changes in the twentieth century pointed to the advent of an information age (Castells 1996; Terranova 2004), information society (Bell 1973; Lyon 1988), and information politics (Dean et al. 2013; Dutton 1999; Jordan 2015). Within these debates, the network society, informational capitalism and cybernetic capitalism have also become ways of articulating the driving economic and social forces of the twenty-first century (Dyer-Witheford 1999, 2015; Robins and Webster 1988). In UK policy and research funding the term 'digital economy' is used to designate this attention to the digital as an economic centre. These designations, both in academic debates and government policy, put information and communication technologies at the heart of both social institutions and everyday life.

The concentration across these areas has seen an intensification through the recent dominance of data as a centrally organizing discourse. The language of big data and the agency attributed to data through the idea that things can be data driven have recently taken over whole areas of knowledge production and politics (Cote et al. 2016; Kitchin 2014). This has an attendant data-visualization industry, as well as an infra-structure of databases and interfaces with data, which, as Helen Kennedy (2015) argues in the case of data visualization, 'is a way of thinking that produces numbers as standards and norms'. Understanding life through numbers has a long history, but the intensification of data in multiple forms as a central mode of legibility points to the particular ways that mediation not only entails more media but also changes the conditions by which we understand the world. To see things as data is another way of seeing the world in terms of objects.

In the context of suspicions about representational and textual forms, then, informational representations appear to offer more reliable object orientated alternatives. They offer the illusion of the object, a fantasy of indexicality, where a direct connection to reality is claimed. Like object orientated theories, speculative realisms and accelerationism, the data industry offers an approach to the world in terms of objects. This chimes with the demand that we turn away from meaning making and towards the real by seeming to offer a mechanism by which the real could be apprehended. For example, the production of information about activity through the use of biosensors and the reproduction of this information as a media interface create an informational form of representation (explored further in Chapter 3). Such forms of representation can be aggregated, collected into large-scale data, as well as dividuated into units. These forms of data expression seem to offer a better – as in more real – account of actual behaviours than self-reporting, for example. Such sites of informational media production have become privileged ways of knowing. Media texts created through informational modes of representation, as opposed to those produced through representations, resemblances, composition or creativity, are in the ascendant. Big data, info graphics, code, sequences and algorithms appear to afford different, more material, less symbolic means by which to understand the world. If knowledge has been understood through forms of seeing, and visual culture taken as a privileged way of knowing, then seeing data has become the new visual.

These informatic objects are also constructed through the lens of ideology, situation and perspective, and there is an extensive feminist critique of the rise of new claims to objectivity in data production, collection and visualization (Gitleman 2013; Kennedy 2015; Gregg 2015). However, their appeal is that they appear to offer, at least temporarily, a more robust way of knowing about the world and they require new forms of expertise both to compile and interpret them. In this context data visualizations have become seductive (Gregg 2015), data journalism has become the new journalism, data blasting the new form of scientific hypothesis, and big data a kind of ruling idea that permeates many sectors from science and technology to art and cultural production. The rise of data is part of the consolidation of digital media and its corollary, information politics, as a central mode of contemporary life in western societies.

Data representations, which play out in different ways in the chapters that follow, are new forms of possible digital barricades. They can be used to bring social issues and politics to the fore (for example the 'Schools of Shame' visualization that maps US campuses with sexual violence issues, or 'Observing the 80s' which offers visualizations of the politics of the period). However, they are more often used to create new forms of knowledge in which the processes of meaning making and the decisions informing the expression are locked down and opaque. The idea that big data can shape how we understand the world, from precision medicine to social media analytics, underpins much of the knowledge politics of the moment. This comes together with the rise of public relations and advertising in science and technology, the pervasiveness of mediatization, and the blurring of boundaries between genre, form and regulation. Storytelling about lives, politics and the world draws on data, biotechnology, documentation, metadata and informational forms to such an extent that it seems to put people at a distance from knowledge making, unless they work in the data visualization, data analytics, public relations, informatics or bioinformatics industries. The rise of big data is both a new site of division and inequality and a new site for activism. Digital divides and disconnects are points at which inequality is created or exacerbated by digital and information technologies. Digital barricades are sites of defence, disruption and obstruction, usually associated with demonstrations and activist politics. Each of the chapters that follow identify new sites of division along with new barricades, detailed through stories about technology which in their crafting also aim to make an intervention.

2

The Shadow of Genomics

This chapter explores the UK's 100,000 Genomes Project: Genomics England. The project emerged in the UK in a period characterized by discourses of austerity on the one hand and big data on the other. As such it brings together some of the contradictory choreography of objects and new media materialities. Genomics England is indicative of the accelerated scale of new kinds of data sets for genomics, in which 100K genomes is the new version of a worthwhile sample.

The point of looking at the genome as a kind of unreal object is to try to navigate a path between the way it is given as an object, to which the only response is consumption and use, and the negotiation of it as mediation. Ethnographic studies of scientific work in genomics take the reality of the genome either as a given or as irrelevant. The practices of the scientists, the relations between actors, and the production of meaning making are what is at stake. Media analysis of genomics, on the other hand, takes the genome as a given but asks how it is framed, constructed and made meaningful. Some versions of materialism take the genome as a given in order to provide a scientific discourse about how we apprehend the world (Morton 2013). Taking the genome as a media object enables a new reading and writing of it which disorientates it as an object.

In the 1990s the Human Genome project was the largest endeavour of its kind and the sequencing of a single human genome was the biggest informatic challenge in biology. This focus of resources transformed areas of biology into bioinformatics – a structural change which has yet to catch up with itself (Thacker 2004; Mackenzie 2010; Garcia-Sancho 2012). This change also put the generation of sequence data at the heart of genomics, and the effort to generate multiple genomic sequences from different people, populations and organisms has become open ended.

Politicians, scientists, journalists and publics in the UK have been enrolled in the project of producing genomic sequence data, and Genomics England has been allocated significant resources from the

NHS budget. Genomics is full of promise (Fortun 2008), and has been since its inception in the 1980s, although at the same time its ability to deliver improvements in health care has been limited (Hopkins et al. 2007; Martin et al. 2006). UK politicians have been persuaded that genomics is worth considerable investment and that the UK needs to have its own national stake, hence the launch of Genomics England. Thus, genomics has become an active agent in shaping budgetary decisions, staffing and patient groups in relation to the NHS.

Genomics has become a central factor in how people understand genetics, identity, evolution, health and behaviour. The double helix, DNA, the gene and genome sequences have become central to cultural, biological and medical narratives and explanations about the world. However, although there is a media culture of genomics, and there has been much media production and public engagement in this area, there has been very little opportunity to question whether the whole project of genome sequencing is the best use of resources.

GENOMICS RISING

Discourses of genetics and genomics have seen a wide proliferation and have carried political influence over much of the last 120 years. Evelyn Fox Keller (2002) referred to the twentieth century as the century of the gene, tracing a path from Mendel's experiments in the late nineteenth century, through the social eugenics of the early twentieth century, to the new genetics of the 1950s and the emergence of genomics from the 1980s onwards. Judith Roof (2007) also traces the cultural significance and dominance of DNA in her argument that there is a 'poetics of DNA'. Her analysis builds on early sociological studies of the social and cultural influence of DNA and genes. Since the start of the Human Genome Project a range of studies has emerged, providing different histories and genealogies of the scientific projects and actors involved (Cook-Deegan 1994), and of their politics (Rose and Rose 2014; Reardon 2005, 2017; Tallbear 2013; Tutton 2016). Such politics extend from questions about economics and political economy (Fortun 2008; Parry 2004), to issues of race and identity (Reardon 2005; Tallbear 2013; Kerr and Shakespeare 2002), to those of the constitution of governance and the state (Rose 2001; Rabinow 1999; Jasanoff 2005; Reardon 2017). The historical interest in genomics as a means of testing how much damage nuclear testing and chemical warfare might cause to species stability gets lost in contemporary accounts.

Genomics has become a bioinformatics version of biology (Franklin 2006; Kay 2000; O'Riordan 2010) which, as Adrian Mackenzie puts it, tends 'to move bodies towards a metastable state in which they become more susceptible to different determinations'. Mackenzie also points to the way in which sequences on their own are not relevant in bioinformatics: 'Reading the sequence itself turns out to be far less important than reading the sequence alongside other sequences' (Mackenzie 2010: 321). The size and range of sequence databases is a quickly shifting terrain. Genomes are relational not only to other genome sequences but to other forms of sequence, and there has been a multiplication of big sequencing-type projects. For example, the project to sequence the human microbiome aims to sequence the genomes of the 10,000 micro organisms inhabiting the human digestive system. Such expansions across species, disease and database have created an increase of omics discourses cascading off the designation genomics itself. Thus, discourses of genetics and genomics have shifted in register in the early twenty-first century from things to objects, and now include post-genomics, epigenetics and a language of omics, from proteomics (McNally and Glasner 2007) to the microbiome.[3]

Genomics and its associations have emerged in tandem with infrastructures of information and computational or digital culture. Genomes are manifest as sequences, themselves digital artefacts, existing as sequence data, constituted through elaborate sequencing processes, demanding huge storage capacity. The history of genomics is inseparable from that of computing infrastructures, computational power, speed and storage (Garcia-Sancho 2012). It has also been cast in a similar trajectory: from emerging technology to consumer technology. An editorial in 2000 in *Scientific American* articulated this explicitly: 'Like computing, genetic science is evolving into a consumer technology' (Rennie 2000: 6). The shift in register from genetics to genomics to plural omics is part of this digital ontology, understood in the contemporary idiom of big data.

Genomics and other omics projects are big data projects at the limit, driving capacity and instruments markets based around sequencing machines, chips and data-storage innovation. The vision of an infinite

---

3 Epigenetics refers to changes in gene expression, rather than the genome. Omics has become an organizing term (a bit like -ology) that indicates the study of a field, usually in relation to molecules.

capacity for information storage and manipulation, offered through the promises of computational technologies, unfolds into an infinite demand for genome sequences. Jenny Reardon (2014a) eloquently argues that genomics is driven by the appetite of machines for more data, and poses the question, what comes after? The machines appear to come after genomic data in the sense that they are in pursuit of it. If this machinic appetite is a characteristic of the post-genomic era, what comes after that?

This question provides a jumping-off point for thinking about Genomics England. In 2004 a cluster of newspaper articles about personal genome sequencing appeared in the UK media. These were attached to the public relations work of genome-sequencing companies, orchestrated through a number of high-profile projects engaging a technologically savvy culture of celebrity – including Esther Dyson, Steven Pinker, George Church, Craig Venter, and numerous journalists and media figures of many kinds. A UK documentary about genome sequencing called *The Killer in Me* was screened in the same time period, promoting a Harley Street company which also tried to connect to the personal sequencing vogue (O'Riordan 2010).

At the heart of these projects was the promotion of an emerging market in genome-sequencing services and competition for lucrative contracts in a new era of biomedical projects, marked by the shift from the single Human Genome Project to a number of 10,000 and then 100,000 genome projects. Illumina is a key player in this market, and supplies the sequencing machines for the world's largest genome-sequencing facility, the Beijing Genetics Institute, as well as multiple projects and facilities world-wide. Illumina has been in the business of first microarray chips and then sequencing technologies since 1998. Since 2009, it has offered its own direct personal genome sequencing service, which has decreased in price over that time from $48K to $4K in 2016. As of 2013 the company was estimated to have a 70 per cent share of the international genome-sequencing market (Zimmerman 2014).

As the 100,000 genome-scale projects rolled out in the UK and the US, questions arose about how genomes would be sequenced, stored and analysed. Each of these stages requires substantial and ongoing investment in new technologies, much of which is beyond the resources of single institutions or even single nation states (e.g. Generation Scotland (Reardon 2014b)). There is a concentration and contestation of control over such infrastructure, raising questions about who owns the genomes, where they are stored, who has access to them, and what happens to them in the longer term. Many of these questions have played out before in relation to the Icelandic database (Rose 2001) and to Generation Scotland (Reardon 2014b). In both of these cases data formats became quickly outdated, and questions of ownership, access and flow of information became problematic and contested as both private and public funding became uncertain (Rose 2001; Reardon 2017), the sequencing landscape changed, and the promised benefits failed to materialize.

Illumina provides the sequencing technologies for other national genome projects, like the Icelandic database (DeCode), the Scottish Genomes Partnership (Generation Scotland) and Genomics England, as well as commercial direct-to-consumer (DtC) companies like 23andMe. Illumina's own personal genomics service has functioned as a point of enrolment and an advertising platform for their sequencing technologies. With little medical value, and lots of media attention, DtC sequencing became a big public relations exercise in which Illumina invested by enrolling public figures. In the promotional media culture that built up around this from 2006, DtC sequencing was framed as an elite, expensive opportunity to be part of a new kind of research that would open the door to the circulation of genome sequences and all the projected benefits this would bring. DtC genomics marked a shift in the discourse of human genomics from the single to the multiple genome, but it also became a site of media buzz, promoting attention to and enrolment in personal genomics. In this shift there emerged the promise of democratization and access to genomic information as a new kind of right (Reardon 2012). Direct-to-consumer genomics operated through a discourse of promise, similar to that of the computer in the 1980s. The promise is that consumers will have individual access to the goods of global technoscience and make them their own.

23andMe became the definitive start-up of this era (Reardon 2017); combining social media with bioinformatics, it promised its customers access to their genomes. It was the mid-range corollary to the high-end, elite projects of full genome sequencing. The latter was beyond the price range of all but the richest, but 23andMe and its partial sequence information could be marketed as a Christmas gift for family or friends. It followed the same trajectory as commercial ancestry testing, which a decade earlier had opened up a successful market in hobbyist ancestry DNA interpretation through the National Geographic's Genographic project.

I've yet to have my genome sequenced with 23andMe, but I do have a genomics and me story which plays out a bit differently. In 2009 my mother was diagnosed with a relatively rare single-gene condition that had been identified in the period of the Human Genome Project: spinocerebella ataxia type 6. The medical interest in her genetic condition overshadowed the fact that she also had lung cancer, to the extent that this more fatal condition was not diagnosed until its final stages. Since she died of the lung cancer I don't know how the ataxia would have developed. However, I've also tested positive for the same condition so I may have another opportunity to find out. This family history motivated me to contact Genomics England as a possible research subject. I have twice approached them about having my genome sequenced and been turned down on both occasions. The first time they simply declined to respond. The second attempt involved some correspondence during which they requested details of my relatives. In the end this didn't result in my inclusion because the genetics of my condition are already known. Their response, that the genetics of spinocerebella ataxia are known, is interesting. Nothing can be done about the condition, and as the genetic link is already established it falls beyond the scope of Genomics England. This points to the way that genomics projects are situated on a constantly shifting cutting edge: once gene-disease links are made, the projects move on. Genomes are thus given as an object at the same time as the meaning of the object is subject to an undoing. The question of what the genome is, and what it might do in the world, its reality and meaning as an object, is constantly renegotiated and deferred.

PUBLIC RELATIONS, PROMOTIONAL MEDIA

Illumina courted the great and the good, making genomes valuable objects by offering genome sequencing for the elite. A number of UK journalists and figures with UK media capital, including Steven Pinker, took up Illumina's sequencing promise, and the stories that proliferated became the basis for promotional coverage over a number of media formats between 2004 and 2010 (O'Riordan 2011). Genome sequencing and the promise of DtC genomics, as well as vaguer visions of personalized and precision medicine, became a prominent media topic for a while. In the genome business, however, the imagined consumer includes not only such elite individuals but also institutions, nation states and national health service providers. During this period Illumina successfully negotiated a contract worth £78 million for the UK's own National Health Service 100,000 genomes project: Genomics England.

Genomics England's 2013 launch was announced the previous year with the news that 'Ministers have committed an initial £100m' to the project (Walker 2012). It has been underwritten so far with £20 million directly from the NHS budget, £24 million from the Medical Research Council (both public sources of finance), together with £27 million in investment from the Wellcome Trust, which will establish a sequencing hub at the Cambridge Genome Campus (Wellcome Trust 2014). At the time of writing, ten industry partners have also paid £250,000 each to participate. In the transcript of UK parliamentary proceedings, Hansard, Earl Howe noted:

> The 100,000 genomes initiative, which my Department is funding, is about pump-priming – the sequencing of the genomes of 100,000 NHS patients – with the purpose of translating genomics into the NHS. This capacity will be allocated specifically to cancer, rare diseases and infectious diseases. The service design work will be completed by June and we aim to put contracts in place by April next year. (House of Lords 2013).

Earl Howe was the Parliamentary Under Secretary of State for the Department of Health from May 2010 to 2015. The economic metaphor of pump-priming is explicit here, although the promise of sequencing the genomes of 100,000 patients is slightly misleading, with the patient numbers coming in closer to 75,000. The cancer projects will involve two genomes for each patient, sequencing from a healthy cell and a cancerous cell. Rare-disease patients will have their genome sequenced together with those of two healthy close relatives. In relation to infectious diseases, it is the pathogen itself that will be sequenced. The project involves collecting samples from patients to create a large-scale sequence database, turning patients and diseases into genomes and genomes into objects. It brings public relationships into the centre of the scientific project by enrolling, teaching and communicating through media forms and public relations campaigns.

I referred earlier to Jenny Reardon's observation that it feels like there is something machine-driven about genomics. The development of sequencing machines requires the generation of terabytes of data that no human can handle. An effect of sequencing machines is to demand data. It sometimes feels like human genomes are being produced to feed the appetite of, if not of the machines, then the shareholders of Illumina. These anxieties might seem similar to those that have cast a shadow over industrial production writ large. The fear that human bodies are feeding the machinery of capitalism has been a popular trope from Marx onwards. However, as many others have pointed out (Fortun 2008; Parry 2004; Sunder Rajan 2002; Haraway 1997; Mitchell and Waldby 2006), it is important to account for biotechnology as an industry because of its conflation with biology and the ideology of natural ordering. The sociologist and information theorist Manuel Castells has long argued that biotechnologies are also information technologies (Castells 1989; Parry 2004). The circulation of biocapital, and the conversion of human surplus, or life itself, into commodity forms that can also circulate, is a condition of the economy of genomics. Although the UK retains something of the National Health Service, Illumina has been a significant force in driving the market in genome sequencing in the UK, partly through enrolling actors in this process through promotional outreach. *Forbes* has this analysis of its economic position: 'Since 2008 Illumina's sales and profit have both increased 147%, to $1.42 billion and $125 million, respectively, as the stock increased 617% and the company's market capitalization reached $23 billion' (Herper 2014: 1). The article also draws on other sources to make this observation: 'It's rare that you find a company that has 80% to 90% share of anything and is driving the technology so fast that nobody can catch up' (Herper 2014: 1). This points to the near monopoly status of Illumina in the political economy of sequencing.

When 23andMe opened its services there was much discussion about whether personal genome sequencing was a brief fashion or a viable tech industry start-up. The technology press lauded the company and it was headline news for a while. It initially ran into some issues with the US Food and Drug Administration (FDA), but the roll out of its tests has continued and it now advertises on daytime TV in Canada and on billboards in the UK. Like Genomics England, the company has also put a lot of resources into media production. Its entire interface is web-based; people send off for their test kits and then become enrolled through a website which presents the genome sequence information through a strong interpretative interface. As a web-based company trading in sequence data, 23andMe is media all the way down. It is part of a social media paradigm in which the app, website or platform is the locus of interaction. There has been much ink spilled, in both journalism and academic writing, over the question of whether its service is value for money or whether it gulls consumers into buying worthless information. 23andMe and Illumina were both instrumental in creating the high media profile of personal genome sequencing in the early 2000s. Illumina, however, whose consumers are not the relatively well-off individuals of 23andMe, but elites, nation states and global institutions, has not been subject to the same kind of scrutiny. Social science engagement with genomics has addressed questions of political economy, but there has been no public engagement around the question of the NHS becoming an Illumina customer.

GENOMICS ENGLAND AND MEDIA PRODUCTION

Genomics England has hosted public meetings and engaged in media promotion in the first few years of its launch. It produces two different kinds of videos and animations in-house. The first type are participant testimonials and explanatory pieces about the project or about genomes. The second type are designed to engage patients and wider publics in a public conversation about genomics. The series of participant testimonials consist of talking-head interview formats with participants, including parents of children involved in the project. In each case the narrative sets the participant up as an advocate for the project and for a future of health care in which their data will make a meaningful intervention. The possibility of treatment in the lifetime of the patient is usually disavowed or played down, and the participants emphasize that they have no expectations about this, but they all express the idea that they are providing data that is necessary for future drug development and future possible discoveries and treatments. These videos enable the promotion of the project in the participants' own words, and the display of a range of ages, ethnicities, genders and illnesses provides an open and inviting feel to the material.

A slightly differently pitched series of animations have also been made that are focused on specific issues, and these have more didactic qualities. These are explicitly positioned as public engagement materials. This social science and public engagement element of Genomics England is currently titled 'Socialising the Genome'. It is led by Dr Anna Middleton, who articulates the relationship between the project and media production as follows:

The project is particularly exciting due to the novel partnership we have set up between social science (me) and the creative advertising world (Julian Borra, Global Creative Strategist and Founder of Thin Air Factory and ex Saatchi and Saatchi Group Creative Director). Julian and I are using our collective skills to see if we can create a 'populist, scalable conversation'. I provide the material; he provides the razzamatazz. (Middleton 2016)

The outcome of this partnership is a series of animations embedded as YouTube videos, dubbed as 'gene tube'. The animations are called: gene deck shuffle; gnome; reasons to be cheerful; glitch; searchme and dnazing. They each feature stop-animation-style line drawings, such as in the image from 'glitch' in Figure 2.1, which explains that everyone has glitches in their genomes.

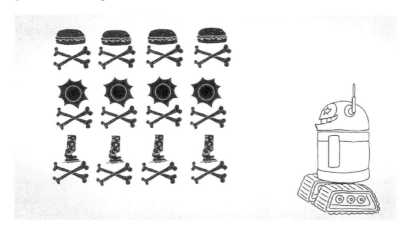

Figure 2.1 Scene from *Socialising the Genome* series of animations: Glitch (Thin Air Factory)

The animations are explicitly designed to try and engage publics in conversations about genomics. They are short, well-produced films that raise, in very general terms, some of the issues derived from Middleton's focus groups. Like much in the field now, they work to reattach the visual culture of genetics, which has a lot of resonance (Roof 2007), to genomics. The same issues raised in social and ethical research around genetics in the 1990s crop up again, such as: questions about how genes are passed from one person to the next; what the implications of research are for insurance; issues about control over data; and the assertion that everyone is equal in that we all have 'glitches' in our genes. The newer turn that didn't feature in earlier iterations is the promise that personalized genetic and precision medicine will be an outcome of current research.

The animations don't anchor their narrative and images with any specific resources. For example, there is no reference to legislation that might protect people from exploitative use of their genome sequence, just an assertion that it won't happen. The question of genetic discrimination is dismissed with a declaration that because we are all equal in having genetic glitches, to attack any one of us is to attack us all, and therefore we have the collective power to declare hands-off in relation to our individual genomic information. Again there is no direct link to legislation or mechanisms that might ensure this. This claim seems slightly ironic coming from an initiative that is all about collecting genomes at scale, and it is interesting to reflect that some of the genomes sequenced are genomes not of humans but of cancers and other diseases. This glossing over of distinctions between human and nonhuman genomes hints at the problem of agency here.

The animations raise issues about understandings of family and reproduction, for example they conflate biological parenting with parenting per se. In 'the gene deck shuffle' piece, sexual reproduction is described as a way of shuffling genes as though they are cards, and the title is suggested as a new more educational euphemism for sex. Genomics-related research has, in theory, unsettled the heteronormative story of the nuclear family through practices such as mitochondrial transfer and therapeutic cloning, and through claims that DNA testing demonstrates that paternity is rather less sure than it looks. In practice, however, clinical genetics and genomics are often used to reaffirm and naturalize the straight story, and Genomics England is no exception.

These animations are made in partnership with Julian Borra's production company, the Thin Air Factory. As short conversation pieces to stimulate discussions about genomics they are potentially quite fun, although they can also alienate through their lack of address to blended, adopted or queer families. They join a well-populated genre of animations representing DNA, genes and genomics, from the Mr DNA science clip in *Jurassic Park* (1997), to the thousands of didactic or promotional clips circulating online currently. However, it is their location as embedded within a biomedical research project and as both explanatory and promotional materials that is significant. Although the use of promotional materials in share prospectuses and in technology

R&D is not new, the positioning of PR materials and advertising as part of the materiality of emerging technologies indicates a shift in the intensity of mediatization and the role of the media.

Despite or probably because of the very high media presence of genetics over the latter part of the twentieth century, the high visibility of Human Genome Project, decades of funding science and art projects, media engagement, numerous Hollywood films taking up the cause, and a whole field of social science literature dedicated to public understanding of genetics and genomics, it would appear that genomics still has a legibility problem. Certainly Genomics England is representing things that way.

One story that Middleton tells from her focus-group work, with people directly affected by genomics, is that people don't think they understand what genomics is. Middleton hints that this might be part of an identity construction in relation to science: 'Socialising the Genome has done a series of focus groups with members of the public to explore what people already understand about DNA and genomics – even if they think they know nothing – and how they are currently talking about it' (Middleton et al. 2015). The formulation here, 'even if they think they know nothing', gestures towards fairly consistent and long-term findings in the social science literature that people often disavow their own scientific knowledge, while also constructing fairly complex understandings of science.

However, at the same time as including this gesture towards possible lay knowledge, Middleton also creates a fairly dismissive framing of these publics by citing them as often confusing genomics and gnomes. This is written into the series of animations, which feature images of garden gnomes. One shows a hand pushing the first letter 'e' out of genome to spell gnome. In this way the visual materials play on the confusion of 'gnome' and 'genome'. However, although the gnome trope has become a familiar character in stories which represent the public understanding of genomics as a deficit, there is no reference to this longer history in Middleton's account.

The gnome/genome trope has been circulating for well over two decades. In 2000, *Scientific American* published a series of articles called 'The Business of the Human Genome'. In one article, staff writer Carol Ezzell wrote: 'what a difference a decade makes. Time was when politicians – not to mention the general public – didn't know a genome from those diminutive forest-dwelling fellows of folklore. In 1989, for instance, President George Bush made a genome-related gaffe in a story I've been dining out on ever since' (Ezzell 2000: 49). Ezzell refers to Bush mistaking 'genome' for 'gnome', and his alleged lack of understanding about science was used in the US media to protest lack of support for the biosciences during his term in office. Ezzell asserted that things were very different now (in 2000), maintaining that few people can say they've never heard of the genome project, and that it had 'burgeoned into a multi-billion industry' (2000: 49). Following on from this trope, one of the leading genomics industry players is called Knome (set up by George Church in 2008), which continues to keep gnomes phonetically in play while also being a portmanteau of 'know' and 'me'.

Since 2000 there have been innumerable research networks, ELSI programs, biomedical initiatives, popular science writings, novels, films, television and art works engaging and promoting genomics. In addition, the broader field of public engagement with science as a whole has been formulated in relation to the idea that publics and lay knowledge making are rich in resources for talking about genomics. In this context, Middleton's reference to ignorant publics and a deficit model of knowledge about genomics seems mistimed.

One explanation for the legibility problem is that genome hype never took quite as easily as gene hype, and that while the gene may be the icon of the twentieth century, the shifting discourse from genes to genomics was unable to build on this, and the new terminology has just confused things. Another issue is that other sciences have entered the frame. While the 1990s seemed dominated by genetics and biotechnology at the forefront of popular science, the early twenty-first century has seen the ascendance of neuroscience and a range of omics projects, increasing competition for public attention and investment.

However, another important structural factor is that building a case for anxiety about public understanding enables funding and support, and genetics and now genomics have been repeatedly framed by experts as misunderstood by publics. The reinvention of this claim facilitates a kind of ground-zero approach where the history and context of all prior research and funding can be bracketed out. Hence Genomics England's ability to enter the field without acknowledging any connection to previous research in this area. This kind of amnesia is expensive, however, requiring the repeated injection of resources into the same work over time.

Nevertheless, the claim that an area of science is misunderstood has been an effective mechanism for opening up support. When that area is also represented as cutting edge or vital to biomedicine and the economy, as in the case of Genomics England, it is an especially effective appeal. In this light the reiteration of tropes about ignorant publics mistaking genomes for gnomes can be understood as a rhetorical device for securing a licence to practise. Certainly the media work in the Genomics England project is entirely promotional, and doesn't provide a platform from which to question the goal of sequencing genomes at scale. The mission is 'To improve human well-being by increasing the individual and collective engagement of everyday people with the positive benefits of genome science' (from Borra's LinkedIn profile). There is no mission to facilitate a space in which the project itself could be questioned, and the framing of genomics as complex and unintelligible (without expert media guidance) continues to mystify and obscure the issues at stake. Thus, the making of these objects demands an erasure of history and constructs the genome as an impossibly difficult object that has to be continually preserved and rediscovered as such.

These promotional materials allow genomics to be positioned as already having positive outcomes. However, these outcomes are as yet uncertain. In 2013 23andMe were issued with cease and desist notices by the FDA because they had not been able to classify their personal genome sequences as having proven medical benefit. Peer-reviewed research from Genomics England in the public domain includes the following statement: 'Raw sequence data could, in theory, be returned instead. This might appear nonsensical as, on its own, it is a meaningless code with no clinical value' (Middleton et al. 2015). These tensions between the promise of genomics as an object which can be used to deliver health-care benefits, the difficulty in interpreting the sequence data as meaningful, and the difficulty in making genomics legible to publics (including health-care workers), are part of what is playing out here.

Genomics England has engaged publics in multiple ways, through recruiting patients and medical workers, media coverage, public meetings, questionnaires, surveys and campaigns. Some of its material is produced by a creative advertising agency and public relations and promotional publics have become bound into what was once thought of as upstream research. Thus, genomes gain legitimacy as objects in relation to media engagement, persuasive messaging and a spectacular media culture. Genomics England is as much about public relations, engagement and media making as it is about scientific research and medical benefit. Genomics research has thus seen a shift from public engagement to public relations. In the UK, a particular tradition of engaging publics in science through forms of media production emerged from the *Science and Society* report in 2000 (House of Lords Select Committee on Science and Technology 2000) and later reports about public understanding and trust. These reports could be summed up as claiming that there was a lack of public trust in science and advocating much greater communication with publics about science. They shaped subsequent research and engagement activity such as the art-science funding stream from the Wellcome Trust. Over time this has engendered a proactive promotional science media culture, including celebrity scientists and (in the UK) the Science Media Centre (Haran 2011a and 2011b).

In 2007 the Wellcome Trust published a collection called *Engaging Science: Thoughts, Deeds, Analysis and Actions*. This summed up and reviewed the previous decade of public engagement activity around biomedical research. Much of this was concerned with developments in genomics; for example, the exhibition Generation Genome toured British science museums and centres from 2007. An earlier initiative was the genomic portrait of the biologist John Sulston by the British artist Marc Quinn, commissioned by the Wellcome and launched in the National Portrait Gallery in 2001. Around 2000 a number of exhibitions were commissioned, by the Wellcome in the UK, and by the American Museum of National History in the US, and at the time the extent of this kind of public engagement created concerns about hype and over-selling. Ten years later another round of engagement activities celebrated the ten-year anniversary of the first draft of the genome. In the meantime television documentary and drama productions were driven by genomic themes, for example *If Cloning Could Cure Us* (2004) and *The Killer in Me* (2007). This has created a high media presence for genomics to the extent that it has become not just an icon or a poetics but a pervasive and influential discursive formation which shores up beliefs in things such as genetic determinism, gay genes and master genes.

There have been a number of concerns expressed that this media engagement amounts to constructing the phenomena it purports to discover (Stevens 2008), and accusations of corporate sponsorship and genohype have been a feature of this engagement culture (Catts and Zurr 2005). However, these issues have become backgrounded and media cultures of genomics have become less critical over time. As genomes have become pervasive and naturalized as objects, critical attention to genohype has also moved on.

The turn to publics has given way to a public relations culture in which promotional media collaborations are used to roll out genomic research rather than provide any opportunity for questioning whether this is in the public interest. In her analysis of the public relations role of the Science Media Centre in the UK, Joan Haran (2011a) points to the way that the media becomes the grounds of governance. Because this engagement occurs at the research stage and as part of the research process, public relations materials, advertising and their audiences themselves become part of the research materials. This is part of the travel between science and technology, between knowledge and commodity forms, as materials and capital travel through different nodes of communicative capital and biocapital. It is also part of the ways in which things are made through mediation, or in the media, but are also taken as given objects such that a promotional film can be (mis)taken as forming part of the material base of an emerging technology.

Genomes can only be detected through complex scientific apparatus which produce the textual object: the genome sequence. However, the sequence is still not an immediately legible object and is referred to as raw sequence data. This requires further layers of documentation, transferring the sequence from one format to another until it is 'readable'. Even when readable as a sequence of nucleotide acids, however, it still doesn't have meaning as such. Thus, the attachment of interpretative interfaces, and storytelling materials about what sequences mean, also become an integral part of its construction as an object.

TALKING, READING, WRITING AND SILENT GENOMES

The Human Genome Project and the proliferation of such projects since have contributed to the visual culture of genetics and our whole way of talking about genes. The extent to which genomes have become a technology of story is very widespread, as I have emphasized in relation to the role of media making in the Genomics England project. In contrast, a closed meeting at Harvard on 10 May 2016, focusing on the project of synthesizing the human genome, garnered much media coverage because it deliberately sought not to engage the media, and the invitation-only list of attendees was a narrow one. Several prominent invitees declined to attend because of this elitism and secrecy (e.g. Drew

Endy, Jeremy Minshull). However, the participants lack of openness didn't prevent media coverage of the meeting. The construction of the meeting as secret provided an additional storyline, exacerbating the coverage of the meeting as an event. The participants published their proposal for synthesizing the human genome in *Science* a couple of weeks later, and the controversy around the secrecy of the meeting facilitated the publicity for this proposal.

The proposal to synthesize the genome lays out a ten-year plan for a second Human Genome Project, this time called Human Genome Project-Write. In the proposal the original Human Genome Project (HGP) of the 1990s is renamed the Human Genome Project-Read. The implication is that the HGP enabled a read of the human genome and the synthetic genome project will enable scientists to write genomes. During the 1990s and the early twenty-first century the HGP was described as decoding the human genome, and at its nominal completion it was referred to as God's language, a book and instruction manual, a code and blueprint. President Clinton said that 'we are learning the language in which God created life' (The White House 2000: n.p.). The bioscientist Craig Venter, who was involved in the HGP, references this decoding association in his autobiography *A Life Decoded: My Genome: My Life* (2007). When the first draft of the HGP was announced Venter said it was the 'first time our species can read the chemical letters of its genetic code' (The White House 2000: n.p.). Francis Collins, Director of the US National Institute of Health, announced the 'revelation of the first draft of the human book of life', going on to say that 'Many tasks lie ahead if we are to learn how to speak the language of the genome fluently' (The White House 2000: n.p.). Tuned to the book of life trope, the sequence of the human genome was printed out in book form, and can be read at the Wellcome Collection's Gallery.

However, despite the literary references to reading, writing and language, the book format has no scientific relevance in genomics. The genome is an artefact of digital media, its framing as code has been more dominant and the sequences are what have currency in scientific research. Reading and decoding in literary terms refer to processes of meaning making; in those same terms the first HGP was a process of encoding, that is making a text, not in fact decoding, which would be to interpret a text. Large parts of the HGP still remain meaningless and the question of how to make them meaningful has been a driver in genome projects ever since. The reference to reading in the first iteration of the HGP concerns a computational kind of reading, retrieving and acquiring data one unit at a time and assembling it in sequential order. This mode of read-write is reiterated by the recasting of the original HGP as HGP-Read and the casting of the new proposal as HGP-Write. Writing doesn't have a specific valence in computing: to write usually means to be able to write and edit code or encode data. Reading and writing in computing are put together through the read-write designation. Read-write means to be able to read and edit, as opposed to read only.

The HGP-Write proposal then evokes a project in which a synthetic human genome is constructed for editing purposes. The proposal comes on the back of the popularity of, and controversy around, CRISPR-Cas9 (Thompson 2015; Vasiliou et al. 2016). CRISPR is a novel genetic editing technique, established in 2013 with the potential for take up in every kind of genetic editing lab in the world (Thompson 2015). The acronym stands for Clustered Regularly Interspersed Short Palindromic Repeats. As a technology it promises to radically change genetic editing, making it easier and more accurate to edit genomes. This is relevant for gene therapy, or the genetic editing of an organism in its own lifespan, but also for genetic editing of embryonic, reproductive cells and germ-line cells.

In December 2015 a meeting between UK, Chinese and US scientists called for caution in allowing any editing of embryos, or germ-line editing that would be passed onto future generations. Since then, research involving CRISPR-Cas9 and embryos has been licensed in the UK. However, UK regulations prevent implantation and these remain experimental conditions. The combined prospect of genetic editing together with synthetic genomes evokes the conditions of what Craig Venter refers to as digital life: synthesizing and editing genomes. George Church is also a key proponent of CRISPR and a leading figure in the HGP-Write proposal, and like Venter he is also heavily invested in genome editing.

Framing the HGP in terms of a read-write dichotomy reminds us of the instantiation of genomes in computing infrastructures. They are already digital and amenable to the same circulation of information politics (Jordan 2015; Terranova 2004). Genomes as media objects are also temporary stabilizations in processes of mediation. Human subjectivity can be understood as always being with and in media in the sense that communication and mediation are integral to human selves and societies as we understand them (McLuhan 1964; Kember and Zylinksa 2012). In this sense all media are biomedia, part of a process of making life; bodies are shaped (although not determined) by the patterns and meanings in the representation of bodies, through gesture, posture, expression, diet, dress, body modification, drugs and surgery.

Genomes promise to make life differently, to intervene inside bodies, to change and to perhaps make new patterns in life making. However, although genomes offer the promise of the radically new – genetically modified species, new organisms – their effects so far have been to reproduce older patterns which privilege particular bodies. One example of this patterning of conformity is what Tom Shakespeare (1998) refers to as the 'weak eugenics' of individual rights models of genetic testing and screening, which has resulted in a dramatic increase in terminations of pregnancies following positive tests for Down syndrome. Troy Duster's (1990) analysis of the new genetics points to the intersecting processes by which new genetic sciences appear to reproduce the same old social inequalities. In fact they exacerbate them through the intersection of genetic accounts of life with mistaken (or disingenuous) accounts of the transmission of privilege and economic interest. The idea of the HGP-Write opens up the possibly of intervening in the genome as code; this makes it part of media production, an intervention at the level of representing things in order to constitute them. If genomes can be written then it makes sense to look at what happens at the level of representation. For example, there are potentially similar dynamics in the relationship between coercively normative images in the representation of women's bodies (white, thin and able-bodied) and the prejudicial and discriminatory representations of disability and the choices made in relation to genetic screening for Down syndrome.

Genomes contribute to processes of mediation, they are part of making media identities. Media identities, like media objects, are moments of stabilization in processes of living and making lives and meaning. The promise of editing the genomes of organisms, writing the biomedia of life, is also a promise of the capacity to make a cut (Zylinska 2009; Barad 2003) in the world and stabilize the meaning of life in a particular way. This promise is hubristic; all processes of mediation are excessively productive and generate destructive irruptions that make more than their producers encode. However, the hubris of science and technology is more productive of investment, attention and belief than other forms of hubris, and that of the biosciences particularly so. Hopes that diseases can be cured, disasters averted and better worlds made go hand in hand with what Welsh and Wynne (2013) refer to as 'scientism'.

That genes are communicative forms became explicit in the 1990s through the way that genes talk. Gene talk circulates from casual comments about the gene for this or that or the DNA of this and that, to medical diagnoses referring to genetic mistakes, through films about genetic modification. Gene talk is a recognized genre in the history of science and the sociology of health and medicine for looking at the way people (both in the sciences and outside of them) talk about genetics. Evelyn Fox Keller wrote in *The Century of the Gene* that 'at the very moment in which gene-talk has come to so powerfully dominate our biological discourse, the prowess of new analytic techniques in molecular biology and the sheer weight of the findings they have enabled have brought the concept of the gene to the verge of collapse' (2002: 69). Keller argues that gene talk is persuasive and powerful, and that even when specialists disavow it or critique its reductionism the same actors use it and recognize that as a rhetorical force it works well. This in part explains some of the investment in the language of gene talk across multiple sites (Lindee and Nelkin 1995; Stacey 2010; Roof 2007). For example, geneticists often believe that a much simplified version of their own register is required to communicate with patients and publics, and these attitudes about public understanding lend themselves to modes of talking about genetics that are vague and drift towards determinism and agential master gene narratives. This language is replicated in the animations produced by Genomics England, where sexual reproduction is re-cast as the gene deck shuffle, dealing us our genetic hand.

Gene talk and talk of genomics operate in the same discursive universe. Gene talk was fashioned through the Human Genome Project, and although the term genomics puts up barriers to recognition, once this is overcome, genomics talk looks a lot like gene talk. One way that genomics operates, however, is to suggest a break from genes and gene talk. If gene talk had become somewhat discredited in the 1990s for over-simplifying genes or genetic determinism, the term genomics offers an opportunity to say things are more complex now but without really changing anything. Genomics as a term suggests complexity, something a little more difficult. Audiences are assisted in understanding this, for example, by Middleton's helpful animation explaining that genomes are not gnomes. There is always a pause after the question, what is genomics? Reaching for the HGP or the claim that genomes are about whole organism genetics is a way into explaining genomes, but once in, the talk is all about genes. At the same time that genomes signal the collection of hundreds of thousands of genomes and the mystique of bioinformatics, Genomics England continues to tell stories of the gene deck shuffle.

Some genes talk more than others. Gene talk may have peaked at the end of the twentieth century, and the new discourse of genomics is similar but a bit more removed, less popular, more complex. However, there are still some genomes that are more talkative than others.

The Human Genome was a difficult idea in the first place: if a gene is almost impossible to define or know, how can there be a singular human genome. This has been answered by the explosion in genome sequencing, but the idea of a base-line human is still pervasive. Craig Venter's genome was in the pool for the five people whose genomes were the basis for the human genome, hence the second part of the title of his autobiography *A Life Decoded: My Genome: My Life*. In 2007 the first named human genome sequence published was that of James Watson. These specifics, together with the DNA portrait of John Sulston in the UK National Portrait Gallery, represent the presence of the genomes of bioscientists as having a very high incidence among those that have been read.

The human genome sequenced in the HGP is also referred to as the reference genome. The idea of a base line comes together with the kinds of selective patterning described above. The base-line genotype is in tandem with same phenotype of privilege, that of an elite white man practising in an elite white male field, reproducing an apparatus of knowledge production composed of scientism and the veneration of the figure of specifically heterosexual and masculine forms of scientific celebrity (McNeil 2007). What then of the base line for the HGP-Write? How will anything outside of what is taken as normal through that apparatus pass into the HGP-Write? Who will get to make the cut in the editing of new genomes? If the promise of life follows the sign of biomedia in making new genetic organisms, it looks a lot like the same crowd.

The 2016 Harvard meeting about the proposal to synthesize the genome garnered a high profile for being secret and exclusive. However, like other forms of secrecy and exclusivity, this also creates value. Exclusive high-tech biotech projects supported by institutions like Harvard and scientists such as George Church are where the money and scientific capital is at. The UK government decided to buy into Illumina's sequencing empire in the hope of securing a share of the spoils, but it is not clear who will directly benefit from this, aside from its boosting the careers of scientists, the profile and revenue of creative agencies, and shareholder value.

The HGP-Read brought fame and fortune to an elite, and improvements in health care to a few. Genetic testing and screening have had more impact on the selection of embryos and abortion decisions than any other area. In the excitement and promotion of genetics, many other areas of health care have been backgrounded or overlooked. For example, my mother's lung cancer was misdiagnosed until it was much too late partly because one of her tests showed that she had a rare genetic condition. The attention to the latter and her referral to a genetic consultant meant that other test information, such as X-rays, were not followed up and the tumour wasn't detected.

Base-line genomes and reference genomes will continue to be sequenced in relation to whatever the dominant version of a good phenotype is. Constructions of race, disability and species will continue to be attached to genomes and condition how they are understood, although the question of how to interpret the genome remains elusive (Reardon 2017). One approach to this has been the big data claims for an hypothesis-free science, the idea that if we collect enough data things will just emerge and the data can be blasted for findings undetermined by human constructions. This seems like a false premise, since data is never raw, and genomes are always already ideological, communicative media objects through which life is made.

MATERIALS: DATA, JUNK AND OBJECTS

The term 'raw data' has become a shorthand for the genomic sequence data, prior to interpretation or reading for meaning. However, as Lisa Gitleman (2013) explains '"Raw data" is an oxymoron.' This is true of genomic sequence data which isn't a naturally occurring flow of signals from the genome but a very carefully constructed read of signals correlated to current understandings of biochemistry, reflecting 50 years of developments in sequencing.

The genome is understood as all the genes of an organism. A genome sequence is a reading of the order of the four bases of DNA, that is G (guanine), C (cytosine), A (adenine) and T (thymine). The raw sequence data is a recording of light refracted through the DNA sample identifying the order of the bases. The encoded light patterns to create a string of text – or a read. This is represented through the familiar lines of letters

GCAT. There are 3 billion of these letters in a human genome, hence my claim that genomics is big data at the edge. Its smallest unit has 3 billion signs. Genomics England hopes to collect, process, store and make meaningful around 70,000 times this 3 billion. This is where big data and highly specialized computing power comes in, and the importance of who controls this part of genomics is crucial.

It is tempting to draw on Bardini's 2016 book *Junkware* at this point. Bardini argues that if we are slaves to the machine or to DNA as a machine, then we must take junk seriously. Starting with a critique of junk as in junk DNA, but following junk though culture, virus and code, Bardini argues that finding that we are junkware opens up new opportunities for breaking out of a system in which humans are the reproductive apparatus for machines. Bardini's writing connects to Reardon's question, what comes after and before machines? But Bardini answers this through the provocation that humans are merely the computer's way of making more computers. This is similar to Dawkins' selfish gene trope, which Bardini also takes up, in which human evolution is merely the mechanism for the reproduction of genes. In both of these formulations the genes and the machines become lively agents and being human is a form of enslavement to these others.

Bardini opens this out to look at the ways in which the category of species is unstable and the human is made up of many different kinds of organism, viruses and microbes. There is more DNA in the many microbes and forms of viral life in the human gut, for example, than in a human genome. Bardini contends that humans can be understood as junkware, both enslaved and radical, through an examination of junk DNA in the register of biology, as manifest in journal articles in the field and through a broader consideration of the cultural meaning of junk across multiple fields from literature to computing. Although these creatively perverse readings of genes and junk open up a critical space around genomics, this critique doesn't make its way into a project like Genomics England in which the collection of sequences is already asserted as a good in itself.

Digital media theorists have examined how algorithms, bots and other software agents build in social patterns of discrimination, ideology, norms and common-sense beliefs, in the same way that other media objects reproduce social norms of beauty, desire or privilege. The same questions need to be asked of genome sequences. Even if the promise of writing or sequencing whole genomes and what this will enable is hubristic, practices of selecting, editing, reading and writing are already with us, biology and computing are thoroughly mixed up, and the airbrushing capacities of Photoshop in creating digital images are not far from the aspirations of genomics in creating organisms.

Timothy Morton writes in *Hyperobjects* (2013) that the world is like an extension of the human phenotype; hyperobjects are just here in our heads, no mediation required. The genome could be thought of as a hyperobject in Morton's terms – his examples are styrofoam and global warming – but the question of what the genome means is not just of interest to cultural studies. It is a primary question in genomics research. It is not obvious what the genome is, and it is not just here in the world with us. It requires a massive apparatus of production to make it appear as an object. The materialist demand that we look at the world as it really is starts to reassemble a universal and transcendent subjectivity in knowledge production. It asserts a view from nowhere by disassembling the imperative to look at how meaning about things is made. Morton, whose writing is closely associated with what has become known as object orientated ontology (or OOO), evokes this perspective from nowhere in the following statement: 'Like God taking a photograph, the non-human sees us, in the white light of its fireball, hotter than the sun' (2013: 50). This invokes what Donna Haraway has called the god trick, a fantasy of seeing the world as though the viewer isn't in the world.

This figuring of inhuman scale is one of the problems with discourses around object orientation, and some versions of materialism and post-humanism: who is telling the story of the nonhuman if not a human subject? And it matters who tells the story. In this version it is Morton who stands outside of the frame even of the god photographer and tells his readers about this enframing. This radical step out of the frame of the located subject is not shared by feminist materialism and it is in this shared aspiration to the epic scale, the reassembly of the universal, or entirely outside point of view, that OOO, aspects of speculative realism, and some versions of digital media theory diverge from new materialism, or traditions of feminist intervention.

A recent special issue of *Cultural Studies Review* on New Materialisms (2015) is a singular example of these two different areas – feminist new materialism, and the digital material turn – being brought together. An article by Asberg, Thiele and Van der Tuin (2015) takes this head on. In their preface to the special issue the editors describe this contribution by Asberg et al:

By performing a critical conversation between feminist materialist genealogies and object-orientated ontology/speculative realism, they argue that the question of feminism in current materialisms boils down to conceiving of immanence in terms of always located, relational and embodied becomings rather than as a new ontological absolute. (Tiainen et al. 2015: 9).

The new ontological absolutes of object orientation try to persuade readers that objects in the world just are. They are immanent, transcendent and powerful beyond the human. In their white light, as Morton has it, human agency becomes beside the point.

Asberg et al. argue that how we perceive things in the world and how we account for the apparatus of perception is still very much to the point. They write in their critique of object orientation and its divergence from important trajectories in feminist and materialist thinking: 'With Haraway we realize that to avoid reproducing the modern god trick of relativism and universalism (transcendence) we have to count ourselves in and stay accountable to our situatedness' (2015: 164). Morton's figure of the hyperobject is seductive and it is a compelling figure to think with, which is why I've included it here. However, it is precisely the seduction of the idea of transcendent objects in whose light humans can only marvel that I seek to disrupt. The genome is not only a transcendent object bringing marvels. It is also a complex assembly of political economies, material infrastructures, data processing, storage, computing power and elite groups of people. It is made meaningful and legible through media production as much as anything else. If it is viewed as a transcendent object it can be used to attract public money and public support and in doing so divert attention away from the politics and economics of the so-called bioeconomy. Calling attention to how it is made through media production, how it drives a big data industry, and how it entrenches the power of already powerful elites is part of calling out its situated and contingent emergence.

3

Biosensory Experiences, Data and the Interfaced Self

The last chapter was concerned in part with questions about health-care priorities, and biosensors have also been significant in discussions about reduced health-care resources. One vision of biosensors is that these new devices for monitoring health and fitness will help reduce demands on hospitals and health professionals. A more extreme vision is that they will revolutionize health care entirely. This chapter examines biosensors through the example of the commercial fitness tracker, Fitbit, looking at a range of experiments with this and other biosensors. Biosensors offer new ways of understanding people and worlds, but they also deliver very mundane and constrained realities. This chapter looks at these devices, their use and promise, to think about a range of different kinds of object-realities as well as what and who is orientated around them. Biosensors perhaps more obviously appear as objects because they are gadgets: watches, Fitbits, Jawbones, Garmins, pedometers, diabetes measurement devices, sleep trackers, cameras, thermometers, carbon dioxide calibrators, and sensors of multiple kinds.

The point of framing Fitbit as an unreal object is to question its inevitability as an object, and to shift the discussion away from studies of use and consumption. Object orientated approaches suggest a world that can be apprehended directly, given training in computation, or induction into the (right) language of theory. Fitbit is in tune with this impulse, promising a direct apprehension of everyday life in terms of units of data. Like the theoretical impulses that the object is in tune with, this is a form of purification. The messiness of texts, meaning making, semiotics, symbolism, fantasy and imagination have an extra-object role. Mediation, and media, fall away in the shadow of the object. Like other biosensors, Fitbit seems to fulfil the promise of an object orientated line of thinking by giving people access to a calibration of the world they would not otherwise be able to access. That is, a perception of the world in terms of units and data, or measured and measurable objects. But neither Fitbit, nor the object world it delivers, are inevitable givens. They are designed choices and could be designed otherwise.

Biosensors offer a kind of networked everyday in which the body is a node in an interactive assemblage of objects and relations, something like a cyborg, but which acts more like a replicant. It is over 30 years since the publication of the 'Cyborg Manifesto' (Haraway, 1985), in which Donna Haraway claimed the cyborg as an ironic feminist figure. Now visions of what once were thought of as cyborg lives propel narratives of consumer electronics. However, unlike Haraway's cyborg there is nothing ironic about fitness tracking or other biosensory regimes. It is quite serious, the interface has become mundane, and the networked human assumed. Earlier worries about blurred machine/human boundaries have shifted to the extent that there is now more concern about not being attached to objects and connected to the network than about being augmented.

Implants, wearables and networks of devices to augment people's everyday lives have become part of the consumer electronics landscape. Apps for tracking running, cycling and walking, or devices to measure movement, heart rate, pulse and blood sugar, come together and produce indefinite amounts of data. Quantification, data collection, data monitoring and big data are features of this landscape. This imaginary of networked, data-producing people and devices is currently manifest in a growing market of consumer electronics. These devices cover a spectrum from health to leisure and blur the distinctions between medical instrument and consumer device. The DIY health-promotion consumer is fast becoming networked into a grid of devices, interfaces and platforms for data generation, collection and sharing. In an extension of the social media paradigm these devices shift the mode of participation from that of creating profiles and uploading information, to that of creating the conditions for automated and indefinite data generation.

One way of framing fitness tracking is to think about how young, affluent, white, female subjectivity has been constructed and constrained over the last 300 years. In this chapter I put Fitbit together with other modes of recording movement and well-being, namely journals and letters from the nineteenth and twentieth centuries, as well as art practice in the twenty-first century. Putting these materials together demonstrates the constraints of Fitbit and other similar devices. Although promising new forms of empowerment, they also act to contain and constrain. The field of activity they encourage is marked out by a limited circuit of steps at work/home and exercise in the gym or in an urban environment.

This contrasts with attempts to hack into, play with and create new experiences through biosensing; these more experimental forms will be contrasted with Fitbit in the following.

BIOSENSORS ON A BROAD SCALE

The term 'biosensors' points to a range of devices already in use, as well as to future imaginaries. The range of sensors currently in use extends across several categories, including highly regulated and advanced medical instruments in biomedicine, the military and sports science; consumer goods that are either regulated as medical instruments in the case of diabetes management, or sold as leisure goods in the case of commercial devices like Fitbit; and physical computing/home electronics goods like Arduino sensor sets. The categories of consumer goods on the one hand and medical instruments on the other are blurred in the current take up of sensors across health concerns.

Biosensing can refer to the detection of biological elements such as pathogens or environmental hazards. However, the term is often used in the current policy climates around health to refer to the detection of biological signals from the human body in order to make some kind of health assessment about individuals (Mort et al. 2009; Lupton 2013). Biosensors fit easily into a health-care environment of which Maggie Mort and colleagues write: 'For almost two decades, health policy in Britain has reflected a conflation of medicine with information and information with modernization' (Mort et al. 2009: 10). Biosensors are also orientated towards nonhuman biosignals including environment, temperature, pollution and sound. They tie in with the idea of the Internet of Things (IoT), in which the internet is understood as more populated by devices communicating with each other than with people. This chapter is concerned with the emergence of wearable biosensors more broadly, but it is grounded in a specific focus on one of the more popular examples of actual take up: the Fitbit brand.

Wearable sensor technologies intersect with surveillance and biometrics (measurements of the body), which could include facial recognition, body temperature and perspiration levels. However, biometrics, unlike wearables, have been cast as impersonal surveillance technologies. They have also attracted explicit criticism for the way in which they objectify bodies and reduce understandings of the body to a limited set of biometric indexes (Magnet 2011). They lend themselves to an object orientated approach via their own processes of objectification.

By contrast, Fitbit wearables and the like, while proliferations of object making, are represented as positively charged devices which offer improved health and empowerment. They are central to what Deborah Lupton refers to as a 'data utopian discourse on the possibilities and potential of big data, metricisation and algorithmic calculation for healthcare' (2013: 14). This data utopian discourse operates across multiple domains, from the technology press to cultural institutions, health care and policy making. Wearables are attached to imaginaries in which surveillance and self-surveillance offer agency and technological augmentation (Mann et al. 2003) rather than the dystopia of a surveillance state.

The surveillance theorist David Lyon has long argued that there are two faces of surveillance: 'The processes that seem to constrain us simultaneously allow us to participate in society ... The electronic eye may blink benignly' (Lyon 1994: ix). Lyon's work traces, at least for the main part, the rise of visual cultures of surveillance: cameras, CCTV, reality television and so on. Biosensors offer a different mode of watching over. Tracking data and signals constitute data selves in a kind of informatic vision.

Wearables as consumer objects are interestingly didactic devices that teach people how to use them and the interfaces they serve, if they have the time (Fotopoulou and O'Riordan 2016). For those for whom building in a training schedule is a part of everyday life they offer a short-cut. Fitness tracking for example is popular among personal trainers. Devices like Fitbit or Jawbone and practices which monitor health also are becoming increasingly common in high-tech employment sectors in which there are incentives for using them.

However, for many they offer the prospect of more labour, and while promising greater efficiency and a better life, also pose the challenge of finding more time to figure them out. They are part of a fantasy that the next generation of technology will make your life work better, and improve your work-life balance, but in practice there is often no time to integrate new devices or get to know them well enough. Such devices often become time sinks and new nodes of informational labour while promising efficiencies and improvements. It is also important to register that they are more toxic waste in the making, even if this aspect is concealed through the dominant message of compliant biocitizenship conveyed through the image of the fit and active young consumer of these goods.

Taking these elements together enables different stories about materiality and embodiment in relation to data. The language of data, platforms, protocols, interfaces, quantification and data visualization has become widespread in attempts to articulate digital transformations in many areas. Key to this is the relationship between the self, identity and agency on the one hand and that of data infrastructures, repositories and institutions on the other.

BIOSENSORY INTERFACES

The question of how biosensors are designed to interface with their users and other devices brings into focus the question of the relationship between structure and agency in new ways. This is an old question, but the problem of how to be an actor in a world in which the terms of engagement are more and more concealed becomes increasingly urgent the more that power is produced through that concealment.

Tim Jordan argued in the late 1990s that a new kind of power came with the information society (Jordan 1999). He saw indications that a new infocratic or technocratic elite would become powerful if the conditions of the social and political world became those of information. In his most recent book, *Information Politics* (2015), he argues that new modes of power are produced as well as new elites. He identifies these information-politic modes as recursion, devices and network protocols. Jordan's analysis has an eye to questions of social justice, and identifies new forms of activism and resistance in these modes of power. Following Alex Galloway's (2004) formulation of protocol politics, Jordan points to the way that specific informatic infrastructures reshape the conditions of engagement with the world.

Interfaces both represent and enact these conditions. Individuals are offered access into the worlds of information politics and infrastructure, with their illusory promises of agency through interfaces with digital systems. Interfaces are something like the obligatory passing points (Callon 1986) of digital culture, along with devices from desktop and laptop computers to tablets, phones and Fitbit wristbands. They are also the point of inscription and data generation. Engaging in Fitbit means engaging with a specific interface that positions the user as aspiring to conform to a healthy template of activity in relation to food, water and sleep. The Fitbit interface can be looked at in two ways. Firstly, the dashboard where information is collected and reused to form colourful and purposeful diagrams.

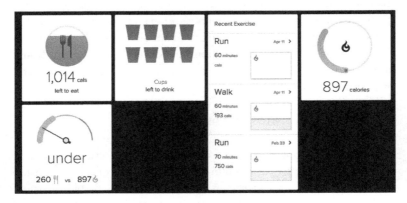

Figure 3.1   Image of the Fitbit interface

Figure 3.1 shows one view of the Fitbit user interface, as that term is usually understood. This is the point at which people are offered diagrams reflecting back the data generated by their own activity or input. The language of the dashboard used by Fitbit has become a convention of the genre, and connects to the use of the term 'data driven' in the big data discourse. In the data-driven imaginary it is as though we are driving vehicles fuelled by our own data, ensuring a productive passage through life, informed by the information coming up on the dashboard of our vehicle. In this configuration, technological platforms become like cars, on the one hand units of driver-controlled power which allow forms of freedom and empowerment; on the other hand units of containment in systems of control, polluting and destructive. A news story about the proposal by the French government in 2013 to try to enforce a tax on data-collection practices by digital industries operating in France indicated this framing in the following claim: 'Your body isn't a temple it's a data factory emitting digital exhaust' (Mahdawi 2013).

The language of 'data-driven fitness' appears in the Fitbit promotional material; the device's dashboard invites agency, but begs the question of who is driving. Offering users a dashboard appears to put people in the driving seat but it feels rather like being a passenger, safely strapped in and well away from the controls, and certainly a long way away from what's under the hood. Like other devices in this market, Fitbit sensors are very difficult to get inside. They are locked up both physically and in programming terms. Breaking them open to access the data in alternative formats is a significant challenge. Because of this, to stay with the car analogy, if you wanted to get under the hood it would be easier to make your own sensor than try to hack a Fitbit device.

In the first hype cycles in 2012 and 2013, when big data was the buzz term in the technology sector, journalists and industry insiders tended to see data as doing the driving (Kitchin 2014: 67). The language that frames fitness tracking offers to put the user back in the driving seat. This slippage around agency is a feature of technology discourses which offer technology itself as an agent of positive change at the same time as obfuscating the ways in which technologies delegate or shift responsibility. Data-driven fitness, for example, obscures the agency beyond the data (commercial data collection), and grounds data as part of personal agency with the fitter human user as a beneficiary. This obfuscation of the industry, design and economic drivers of a device industry naturalizes and reduces Fitbit to the person, the device and their diagrams as though these appeared from the ground of digital culture rather than being commercial interests in large data sets.

It is worth thinking about the Fitbit interface in a second more expanded way. While the dashboard is the main point at which activity is rendered communicable, a second interface is the address constructed through advertising. Fitbit addresses people through advertising, magazines, health and leisure media and the technology press, among other sites.

The interface traditionally refers to the site at which the user engages the computer, and vice versa. This has a history, passing through punch cards, tape, arcade machines, consoles, command prompts and GUIs (Laurel 1990; Johnson 1999). However, as Alexander Galloway points out, interfaces are more processes than objects, and 'to mediate is really to interface' (2012: 10). The two terms are hard to separate. To mediate is to offer a point of communication between agents or phenomena, and to interface is to do the same. The latter term rather more specifically denotes communication between faces, to which, at least read through Levinas, sentience and feeling is attached. However, the interface is used more broadly as a term to mean 'between two things' or at the place in which they meet. I am interested not so much in extending the term interface to mean any point of communication, but in staying with it as a way of understanding computerized devices as sites of mediation. With this purpose in mind it is worth taking the image in Figure 3.2, from a promotional film made by Fitbit about their product, together with the dashboard above.

Figure 3.2   Image from Fitbit advertisement film

In this image the audience is shown an image of a young, white, professional, affluent woman wearing a Fitbit wristband. Fitbit's preferred subject is this kind of young woman. The device is the pink and silver bracelet on her left wrist. The image of this woman and her floating disembodied diagram as she walks down the stairs, on her way out of the house to work, is also part of the interface. Fitbit gathers people into this version of themselves and the technology, in a pairing of unobtrusive, tasteful, ethereal technological sophistication with cultural affluence, privilege and sophistication.

In this vision the actual and virtual are layered together. This conjures a game world which characters navigate with floating vital signs tabs, but also augments reality with a device pairing, in a similar manner to *Pokémon Go*. The rest of the film narrative shows the female Fitbit subject working on a laptop in a café, with the device reminding her to get up and take a few more steps when she has been sitting in front of her computer for too long. Later she goes running, and in the evening goes out with her equally desirable male partner, shown wearing a more heteronormatively appropriate model of Fitbit (chunky and grey), in a more corporate work environment, and engaged in more rugged sports activities. Their going out in the evening is framed as a reward for the productive activity goals achieved during the day as logged by Fitbit.

The images from the promotional film are part of the same address to the subject together with the device and the dashboard. The video on its own is an inscription of a very conventional politics of representation and of disciplinary, normative images. These are also folded into the horizon of the data imaginary and the life that Fitbit offers. Identification with Fitbit means inscribing a personal version of the Fitbit user through the dashboard, as Fitbit recursively takes in data and reflects it back though a personalized profile.

FITBIT AS BIOSENSING

Biosensing is a mode of data generation which connects to both big data and the Internet of Things. These are two closely interconnected discourses that frame communications technology. Big data is everywhere, we might say; as a discourse it emerged from economics and computer science in the mid to late 1990s (Kitchin 2014). As an industry it has emerged in recent years as a reference point for all aspects of social life from politics to science (boyd and Crawford 2012).

Rob Kitchin distinguishes big data from earlier deployments of large data sets (population, climate) through intensification of scale and through the ability to store and read data in relation to volume, velocity and variety. He suggests that prior to big data it was only possible to have data work over two of these three attributes at once, whereas now all three can be achieved and this enables four further characteristics of data such that it can also be exhaustive, fine grained, relational and flexible (Kitchin 2014: 68). The chapter on big data in his book *The Data Revolution* explores these seven characteristics of big data which biosensors seem to fit, potentially generating data in indefinite and variable flows of high volume, high velocity and wide variety.

Fitbit has limited capacity in terms of variety because a limited set of sensors are used. However, in terms of volume the device generates a continuous and indefinite data stream of heart rate, steps taken, distance covered, calories burned, floors climbed, time logs, location data and sleep. Tim Jordan argues that 'recursion is one of the key ... processes underlying the inversion at the end of the twentieth century from information scarcity to information abundance' (2015: 44). Fitbit generates information in abundance, and the relation between Fitbit and big data enables just such a recursive dynamic. 'Recursion produces information from information and then reuses and continues this process' (Jordan 2015: 44). Fitbit produces more information derived from those data streams and reuses and continues the process, so recursion increases volume, and velocity. The data displayed on the dashboard is reused though repeated automated operations to produce charts and reports of cumulative data over time. This biosensory reading of the body produces a form of self-monitoring that is made intelligible through automated logs, counts, charts and diagrams.

It is not that I am unable to count my own steps, but I might lose count or my mind might wander off, and if I spent all my energy on tracking my own activity I wouldn't have the capacity to do much else. Some people do execute such detailed recordings of everyday life. For example, the late eighteenth-century diarist Dorothy Wordsworth tracked of the minutiae of everyday life through her diaries and letters, and the early twentieth-century French writer and filmmaker George Perec was fascinated by exhaustive details and lists. These are practices that cross many forms. However, it is in the mode of information (Poster 1990) that this becomes indefinite, indefinitely recursive, and thus indefinitely productive of more information (about more information). Some of the counting that Fitbit does would never be apparent without alternative sensors, for example counting heart rate and monitoring sleep requires externalization and cannot be self-sensed. More medically orientated trackers measure blood sugar levels (for diabetes) or other internal chemical variables, which can only be monitored via instruments. Biosensors then, even in the leisure market, perform a combination of generating sensible information such as numbers of steps and insensible information such as sleep patterns.

FITBIT AND NINETEENTH-CENTURY LETTER WRITING: 300 YEARS OF DISCIPLINING FEMALE SUBJECTS

An alternative approach to media devices is to look at what they have in common with disparate media forms from different times (Crawford et al. 2016). Taking Fitbit away from objectification as a new media object and looking at it as part of a process of mediation makes it possible to draw out different aspects of what is going on with these devices. One point of entry is to look at the journal facility in the Fitbit dashboard. In addition to the automated collection or measurement of data, Fitbit also enables prose-style journal entries as well as manually entered data. These mechanisms invite comparison with other journal forms, while the dashboard evokes ritual practices of note taking and observing everyday life.

Similarities can be seen here with the use of letter writing and journals among middle-class women of empire in the nineteenth and twentieth centuries. Letter writing and journal writing were highly ritualized practices tied to the regulation and disciplining of particular bodies, particularly those of middle-class women, who were expected to write as part of an everyday ritual, a duty of both communication and self-regulation. These practices were well established and had been so for centuries by the later nineteenth century.

Middle-class women had been writing letters since the late middle ages in the UK, with a significant rise in numbers in the eighteenth and nineteenth centuries. The postal service had been strengthened at the start of the nineteenth century by the use of the telegraph and by rail networks. By 1910 over 15 million letters were handled daily by the UK post office. Taking a snapshot of practices across 300 years allows a different kind of cut through these media forms. Letter writing, by middle-class women, to a range of family and friends marked out ritual time in everyday life. A daily habit was expected in much the same way that Fitbit encourages daily rituals of measuring bodily signals, regular data uploads, and making journal entries. Like the use of Fitbit it also signalled an appropriately disciplined subject, participating in a normative regime of personal hygiene and good habits which also maintained a social world.

Fitbit constructs an ideal subject position, engaged both in appropriate activity and in daily rituals of attaching to the device and checking in to the online interface, referred to as the dashboard. The aggregate of signals assembles a text that is both an expression and a form of recording, especially when feelings and reflections are added through journaling. This compares directly to other forms of journaling, but also to letter writing through the capacity to share these entries with others. The Fitbit dashboard allows sharing of any aspect. This can be thought of in relation to nineteenth-century letter writing, as often letters written at the same time were very similar in content but sent to different people, and shared around an audience of family and friends. Thus, an account of the same activity, thoughts, health and external observations would be sent in similar terms to a friend, parent and sibling so that three or four very similar letters would be written in one morning with only the address and framing changing from one letter to the next.

Two examples of journals and letters from earlier centuries help to draw this out. Entries from the journal of Dorothy Wordsworth written at the start of the nineteenth century indicate something of the patterns established in such writing. She was born in 1771 in Cumbria and her journals and letters are in the public domain, largely curated at Dove Cottage in the Lake District, and some of this material can be accessed through Project Gutenberg. An example from a hundred years later is Kate Hain. She was born in 1885, in St Ives, Cornwall, and wrote to her family and friends throughout 1910–11 while traveling. Her letters are privately held by the family, although some materials are available through the St Ives Museum. My own use of Fitbit and a similar device called Jawbone provide an example another century on. The content of the journals and letters also demonstrates materials that can be thought of as differently structured by Fitbit: while letter and journal forms don't necessarily encourage pie charts and percentages, they do encourage lists, records of dates and time, and notes on activity, health and behaviour.

The following first lines of Dorothy Wordsworth's journal entries for January 1798 make the point that self-observation and accounting for activity was a strong element of the genre:

21st. Walked on the hill-tops – a warm day. ...
22nd. Walked through the wood to Holford. ...
23rd. Bright sunshine, went out at 3 o'clock. ...
24th. Walked between half-past three and half-past five. ...
25th. Went to Poole's after tea. ...
26th. Walked upon the hill-tops; followed the sheep tracks till we overlooked the larger coombe ... Set forward before two o'clock. Returned a little after four.
27th. Walked from seven o'clock till half-past eight. ...
28th. Walked only to the mill.

Fitness trackers originated as pedometers, measuring steps taken. This listing of the detail and repetition of walking activity in the diary form resonates with the way activity is logged in fitness tracking in both the ritual of doing the activity and the ritual of recording it (see Figure 3.3).

The genres of diary/journal and letter writing have many divergences from the interfaces of personal tracking, but it is interesting to reflect on some of the connections. Like the practices of writing letters and journals, the use of Fitbit is represented as a form of everyday

Activity History

Date	Activity	Steps	Distance	Duration	Calories
Oct 15, 12:00AM	Walk	N/A	N/A	13:00:00	1,524 cals
Oct 11, 9:00AM	Run	10,845	6.2 miles	10:20:00	4,845 cals
Oct 10, 9:00AM	Run	10,845	6.2 miles	10:20:00	4,845 cals

Figure 3.3   Jawbone interface showing record of activity

self-improvement, warding off idleness and demonstrating moral virtue. Sharing information through Fitbit is represented as giving a good example or encouragement to others, in a way not dissimilar to letter writing. These letters and dairies were also shared texts, shared around circles of family and friends at the time of writing, and later as part of an historical record, in private and public collections, or remediated in published forms.

Dorothy Wordsworth's letters and journals include many examples of lists and are particularly attentive to recording walks and observations of the local landscape. Her vivid descriptions of natural scenes while engaging in walking are very different to the material gathered by fitness trackers. Such annotation can only be added through the prose-style journal entries in the dashboard. However, it is striking that many of the representations of women using Fitbit in the advertising images portray beautiful landscape elements, most often trees. Although the visual imaginary of Fitbit is mainly confined to a life spent at home, at work, or in the gym, running is framed in terms of the natural beauty of hills and trees, similar to the Bay area landscape around San Francisco.

Like the diaries of Dorothy Wordsworth, the letters written by Kate Hain a hundred years later also contain descriptions of activity and landscape. If Wordsworth's letters and diaries reflect the home tour of UK romanticism, Hain's letters record the grand tour of the colonial imaginary. They contain recordings of activities including walking, riding, singing, chores and eating. Activity and health are observed in detail and again walking features heavily, recorded in almost every letter. On 8 February 1911 she writes to her mother in Cornwall from New Zealand: 'We walk for three days across to Milford Sound and three days back. ... It is only ten miles a day, but I believe very rough going.

Figure 3.4   Fitbit advertising image

However, it will be cool and bracing there, and that ought to help us.' To her brother on 11 February 1911, she writes: 'anyway, I haven't felt so fit all the time and feel the least bit of cold; methinks my gore is out of order, so I am feeding on iron at present. I have got fatter, I think.'

This discursive register is different from the health efficiency rhetoric of Fitbit. Although the word fit appears several times in this collection of letters it is more often used to describe well-being than physical fitness, although both uses are employed. Thus, although the discursive formation and imaginaries are very different, there are similarities in the enacting, recording and sharing of activity as part of a normative and moral pattern.

FROM DISCOURSE TO DATA

There are many significant differences between these media forms and their registers: one is the aggregate collection of data and attachment to an imaginary of big data in which individual data is made meaningful over time, and in relation to larger databases. Another is the informational mode of the representation: unlike journals and letters, which are more clearly modes of representation, the informational quality of the Fitbit dashboard and its automated recording of activity obscures its made-up-ness. Another difference is the relation of the activity to context, especially landscape or location. In relation to landscape, Fitbit's records are indifferent to place. The visual materials in their advertising videos show images of location, some of which can be compared to the romanticism of the earlier letters and diaries in the images of landscape and people. However, Fitbit's records as illustrated above are in the abstract, noting length of time, distance, speed and type of activity, but not location, environment or anything of context. They invoke a data imaginary that can make people more active through aggregating records to produce diagrams, charts and comparative statistics. The highly structured interface demands very specific kinds of data entry. These forms of representation are constructed as at once informational and as having agency in themselves, spurring people on to new goals. Within the Fitbit imaginary it is not the possibility of a sublime encounter with nature that goes with a fit and active lifestyle, but the making of activity into objects, numbers achieved, targets set and goals reached.

This indifference to environment is marked when thinking about different kinds of romanticism, and is slightly menacing when thinking about environmental damage. The register of romantic poetry and prose with which Dorothy Wordsworth is associated demonstrates a passionate attachment to landscape and environment and invokes a natural sublime. This tradition of writing has direct links to the observational methods of natural historians and the history of biology, which retained an attachment to the description and observation of the world.

Wordsworth and Hain's writing was a century apart (and Fitbit another century on). Charles Darwin lived through the nineteenth century, which separates Dorothy Wordsworth's and Kate Hain's written materials examined here. Born in 1809, Darwin died in 1882, three years before Kate Hain was born. The conclusions drawn from his observational writings about species helped to define twentieth-century biology. There are arguments that the central belief structures of the information society are the same as those of romanticism (Botting 1999; Thomas 2013; Wertheim 1999). Fred Botting writes in 'Virtual Romanticism' that, 'Though many of the Romantic versions of cyberculture have yet to be realized, the digital revolution is driven by the momentum of a curiously eroticized and poetic imagination, cyberculture construed as a "habitat of the imagination"' (1999: 101). However, the practices of what can be framed as data romanticism operate with a very different set of attachments. Use of Fitbit orientates users towards intense mediated individualism and a treatment of life in terms of countable units, not as a form of access to the sublime, and it eschews the attendant elevation of nature that characterized eighteenth and nineteenth-century romanticism. The introspective direction of the data collected abstracts both the environment and the past, constructing a highly constricted world of efficiency within the confines of work, home and the gym. Activities are represented as highly repetitive and quite limited. Walking is directed towards an urban environment and Fitbit is designed to detect how many floors a user has climbed, thus anticipating subway stairs and office floors, as well as the number of steps walked. In the advertising images, running is more often cast against a natural environment in which trees and hills appear, allowing for a life outside the gym or work, but the dominant construction is that of an urban environment and repetitive movements.

In the early twentieth century, data, as a term in common use, referred to the facts about something, for example a list of details or events, such as births and deaths, or the material in a police statement. In the popular novels of Agatha Christie, born in 1890, the same generation as Kate Hain, data are physical: 'If I can get a list of recent demises in the Parish ... Whom had I better get the data from – the parson?' (1939: 35), and 'Craddock sighed and stretched out his hand for the data on Cedric' (1957: 145). Letters and journals were also repositories of material and discursive data when they included lists and descriptions. In the case of letters such data could be thought of as networked and aggregated through the postal system. The postal system was a widespread and dominant communication technology at the end of the nineteenth century and the development of rail and telegraph had expanded its remit. According to Duncan Campbell Smith (2011), the first UK airmail delivery was carried out in 1911, but Kate Hain's letters would have been carried by boat and train services. Letters then flowed through the same transport networks as people and other goods, primarily on boats and trains, supplemented by roads. These networks enabled the flow of billions of letters which enabled interpretation at the point of delivery, but which shared only location, address, sender and receiver with the network.

Paper, writing and publishing technologies have their own material infrastructure. Biosensor devices introduce another generation of electronic waste in an era when the afterlife of electronic goods has become a major issue in relation to social justice and environmental issues. In Kate Hain's letters she details walking at the Waikato River in New Zealand:

close beside the Waikato River, and all along the banks, there are boiling pools and geysers of different sizes. The curious thing is that all the pools have different minerals, though they are quite close together; the chief minerals are sulphur, soda, alum and iron, and they are in different proportions everywhere. ... There is a little black pool there containing manganese, the only one in the district. (17 January 1911).

At the time of her visit, European mining in New Zealand had been in operation for about 30 to 40 years. Today, it has been going on for a century and a half and has contributed to major environmental and health damage. Although different in composition to the minerals described above, and differently sourced, Fitbit and consumer electronics include these kinds of materials in their composition. Nearly 200 years of colonial and corporate mining, and the rapacious practices of those industries in the contemporary moment, are implicated in the making of these things as objects.

DEVICES AND PROCESSES OF COMMUNICATION AND MEDIATION

Another difference, and one perhaps more significant than context, is how these forms communicate. Letters have been examined as communicative practices in relation to multiple modes of communication. They can be seen as transmitting a message (Shannon and Weaver 1949), as ritual modes of communication (Carey 2009), and as modes of invoking presence (Milne 2013). They have been contrasted with postcards and the telegraph as well as email lists and other forms of communication (Milne 2013). Internet communication modes, such as news groups, email lists, chat rooms, message boards, gaming, blogs, vlogs and multiple kinds of social media apps, have been compared to letters as they seem to share the imperative to invoke presence over distance, and perform a communicative relation. However, Fitbit is not sent to someone in quite the same way. The interface enables sharing of activity with others, and it lends itself to showing other people the results on screen. Leaning over to show others what your wristband does is a way of explaining how it works and why it is meaningful, but this is less about performing presence and more about performing a particular kind of self. With whom does Fitbit speak as a communicative practice?

In his comparative study of nineteenth-century letters and contemporary online multiplayers, Jordan argues that 'communicative practices are material performances that create presence'. This is a way of articulating an approach to communication as the interdependence of the 'three main conceptual foundations of presence, performativity and materiality' (Jordan 2013b: 53). Jordan explores different approaches to presence, performativity and materiality. He considers presence through a discussion of Milne, Derrida, Levinas and Heidegger, taking in both self-presence and face-to-face presence. The answer to the question, with whom does Fitbit speak?, concerns self-presence. Fitbit communicates a self to the self as well as to others, and like diary and journaling practices it involves a construction of self through the communicative practice. Fitbit offers a way of knowing the self differently, offering reassurance that there is a self, and affirming or disrupting self-knowing. The sleep function, for example, records how much sleep was deep and how much was shallow, and whether the sleep was interrupted. One person I observed using the device reported that it had confirmed what she already thought: that she only had a few hours decent sleep a night. Although she spent ten hours in bed one night, she still felt tired the next day and the device confirmed that only two of those hours had been functional sleep.

Such reports of disrupted sleep, the differential qualities of sleep, and the experience of tiredness are supported, authenticated and anchored by Fitbit. It tells people about themselves, and its capacity for storytelling can be worked in relation to different interpretations. When, for example, Fitbit's records don't correspond with experience or self-image, people talk of it failing to count or record accurately, or failing to sync the data. It becomes a talking point in face-to-face communication as well as in online discussion. People show their screens to each other, explain the interface and discuss its merits, failings or their stories about activity, anxiety, weight loss, energy levels and so on.

Fitbit devices and interfaces also communicate with Fitbit the company, as each record of activity is also a message back, and in this mode Fitbit performs a materiality of presence. It aspires to making these communications more meaningful and already makes them mean something for insurers and employers, indicated by the proviso on the privacy policy: 'Fitbit may share or sell aggregated, de-identified data that does not identify you, with partners and the public in a variety of ways' (Fitbit 2014).

A LIFE RECODED

A return to Stuart Hall's (1973) ideas about encoding and decoding, and to thinking about production and consumption more generally, might enable thinking about this exchange in terms of recoding. If a reader of a media text decodes it in relation to dominant/negotiated and resistant readings of the encoded meanings (themselves multivalent) how does reading one's self in encoded forms work? The Fitbit user is in part a producer of the text of the interface, since each of the diagrams on the dashboard is relational to the input. Further affordances like the profile picture and friends, as well as journal entries, also provide content relational to the user. However, unlike media production per se, there is a very limited range of possible interventions that the user can make. Unlike creating a profile, the content is automatically taken in relation to activity (although manual additions are also possible).

This is neither user-generated content nor a blurring of consumer and producer. It is much closer to the kind of adaptations afforded in game environments where players can modify their in-game avatars and the actions on the screen have a direct relation to the way the player inputs information such as movement and negotiating game play. In older terms, it is what used to be called interactive media, where the actions of the audience have a direct input into what plays out on screen, but always in a very constrained way. The possibility for negotiated and resistant meanings in relation to the dashboard would seem even more constrained if the user has already bought into the image. However, the take up of Fitbit is not seamless. Expressions of frustration at incomplete or inaccurate readings and failures to sync are common in reviews. Also, more highly registered than not, is total lack of use, or users quickly tiring of the object after a few outings. Although Fitbit shipped 20 million units in 2015, it only had 9 million registered active users.

Recoding might be a way of thinking about the disconnect between the advertising image of the ideal subject and the dashboard. Through the dashboard the user's input is encoded through a different form to that in which it is gathered, and this form of recoding is very limited. The representation of everyday life as involving eating, burning off calories and sleeping suggests a very bare life, stripped down to survival. Although this offers a fun and desirable way of engaging with exercise, it is also evokes a trace of the dystopian, a life stripped down to basic processes, signs of life harnessed to a technological device.

The existence of the corporate wellness aspect of Fitbit is worth thinking about in terms of this kind of digital inscription on the one hand, and through this in terms of debates around labour, capitalism and digital culture on the other. Although this book does not take a Marxist or classical critical theory approach (at least not as defined by Fuchs (2011) and Berry (2014)), it is informed by scholarship in these areas. Tiziana Terranova's (2004) argument that free labour, or labour freely given, creates the economic value of digital capitalism, and Kylie Jarrett's (2015) work on feminized labour and the digital economy are both helpful in thinking about the circuits and circumscriptions of Fitbit. Fitbit tracks the labour of the body or the labouring body through counting steps taken, distances run, food eaten and so on. It also incorporates new sites of productivity by counting sleep, which it counts in both senses of the word: in terms of measuring it and in terms of counting it as part of a realm of productive activity. It produces the body as productive; captured time is productive. It captures labour freely given (or paid for through gym membership), while also charging the consumer for the giving through the cost of the device and subscriptions. In creating this convergence of consumer labour (buying or paying) and the biological labour of the body, it capitalizes bios though consumption. In this interaction people give themselves to a digital media circuit enabling biomediation, or the putting of life in media. This incorporates subjects and bodies as biomedia and in doing so enters into a version of life made intelligible through the interface, a life of work, home and the gym, made up of repetitive bodily movements, of steps and calories. This is also a life that is countable, and helps to inscribe a data ideology. The means through which Fitbit is a form of digital inscription is tied to what Deborah Lupton refers to as digitized health promotion (2013). This makes it an ideological and disciplinary device in an assemblage in which specific definitions of health are promoted. It offers a normative imperative towards a limited and prescriptive identity.

Jarrett's (2015) formulation of the digital housewife is also relevant in thinking about the gendered work of inscription into digital culture through these forms. Fitbit is marketed to young women as a lifestyle accessory, through health and lifestyle marketing and in the subject positions constructed through the address. Media coverage of the device patterns much more strongly around sections of the press devoted to women's health and lifestyle. The full range of devices that record and measure activity and health-related data points are gendered in similar ways to the broader lifestyle market. That is to say, while both men and women are targeted, women are targeted much more heavily, and through different devices. For example, the Fitbit One shows a flower and the Fitbit Flex is smaller and more colourful and both are usually worn by women in promotional materials, while the Fitbit Charge is chunkier, black or grey, and usually seen on men. However, as Jarrett argues, the figure of the digital housewife is not only about women's consumer labour in the digital economy, it is about the way the labour of participation is much more heavily patterned that way. The digital housewife is an evocative figure to bring to bear on this area because of the way in which these devices can be looked at in relation to disciplining female subjects through walking over the last 300 years.

As noted in relation to the activity log above, Fitbit aims to capture repetitive movements in everyday life and abstracts these into an account of fitness. The colourful dashboard and its recursive, purposeful accounts of the mundane produce a digital account of the self. Jarrett notes that 'behind even the most sophisticated technological marvel lie the material energies of living human beings' (2015: 2). The Fitbit dashboard captures the material energies of living human beings, and reframes these in the order of sophisticated technology. However, the framing is very narrow, constrained and flattened out. At the same time it makes daily life even more mundane through abstraction. One of the ways in which diaries and letters, which also capture daily life through prose and journal entries, differ is that they articulate a life in relation to a world beyond the medium. For example, Wordsworth's entry, 'Walked through the wood to Holford', is an articulation in relation to a specific experience of place. In contrast Fitbit's record, 'October 15, 12pm walk, duration 1hr', is an articulation in relation to the mediated body, enframed and ready to be converted into activity charts and relations to calories consumed.

Dorothy Wordsworth lived her life in relation to a relatively constrained circuit of travel within the West Country and the North of England. Her travel writing reflects what has become known as the home tour, and she lived primarily with her brother. However, her diaries, even when taken only in terms of accounts of walking, reveal a much more open and relational sense of self and world than does the subjectivity offered through Fitbit. Kate Hain's letters home reflect traveling the routes of empire, from the point of view of the privileged and mobile daughter of empire. They reveal a keener sense of self and others, and of wilderness conquered through travel and technology. Fitbit, attached to a more modern subject, enacts a kind of networked self, seemingly offering offer new freedoms in terms of imaginary and interface. However, this everyday of technology appears to be more empty and more constrained, limited to a circuit of self, device, work, gym, eat and sleep (repeat). Rather than the technological marvel capturing the material energies of humans, the technological mundane seems to obscure the marvels of human energies. And not only that, these devices tend to obscure their own marvellous technological possibilities.

ALTERNATIVE POSSIBILITIES:
BIOSENSORY ART PRACTICE AND EXPERIMENTATION

The discourse of big data is not only tied up with consumer and industry practices, it has its avant-garde, early adopters, subcultures, art worlds and forms of resistance. Another way to think about biosensors, and perhaps think more about their marvellous possibilities, is to open them up to play and intervention. One way through this is to review the range of art practices and alternative products that have already emerged in relation to biosensors. For example, NeuroSky, a start-up manufacturing sensors for electrical brain and muscle impulses, represent themselves in the same kind of language as that associated with biosensors more broadly:

> NeuroSky is breaking the boundaries of health and wellness tracking and analysis by enabling a new generation of consumer wearables and mobile devices. Our advanced biosensors, biometric algorithms, reference designs, and Big Data analytics enable leading-edge innovation in mHealth products and services for measuring body and mind performance. (neurosky.com)

However, their products to date have been orientated towards an electronic games and gadgets market, and their version of a technological accessory looks rather different. An example is their partnership with Neurowear, resulting in the Necomimi headband which features cat ears that adjust position according to electrical brain impulses. It is worth registering that again these devices are imagined in relation to female subjectivity. Most of the images circulating of people wearing this headset are of young women and girls. Although not explicitly an art project, the headband received an honorary mention in the interactive art section at Ars Electronica 2013 (Figure 3.5).

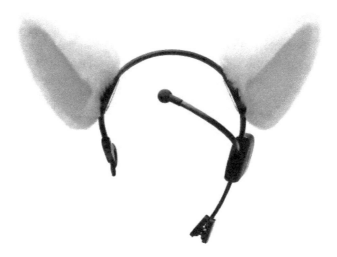

Figure 3.5 Neocomini headset with ears which move in response to electrical brain impulses

In digital art practice Rafael Lozano-Hemmer's *Pulse Room* (2006) and George Zisiadis's *Pulse of the City* (2012) both use heart-rate sensors to express lights and sound respectively. *Heart Chamber Orchestra – Pixelache* (2006) also used heart-rate monitoring to create a musical score, from which a live performance of classical instruments was derived. In *Electrode* (2011) Dani Ploeger used EMG sensors to register the activity of his sphincter muscle and used these patterns to generate digital sound synthesis. Biosensory data, in live and logged forms, has been used extensively across the field of digital art practice. In many ways the use of this kind of data looks very similar to the use of interactive features which have been central to digital art. As the term interactive indicates, the movement of the viewer, audience or user has been incorporated into digital art since its inception. Although inter-activity is a feature of a much longer history of art, it is associated with 1970s video art and most heavily with the use of computer interactivity from the 1990s onwards. If one takes movement as a basic biosensory input then much of the history of interactive art could be opened up to a discussion of biosensory art, but this is beyond the scope of this chapter.

Biosensors have so far had the most commercial success as fitness applications and as medical instruments. In art practice they have been used primarily to collect sensory input from humans, plants, water and air to generate other forms of expression (lights, music), or to create systems that register or make visible phenomena that are not sensible through other means. In other words, biosensors have been understood as opening up the possibility of new forms of communication, whether sensory or nonhuman, for example registering the signals of the body in new ways or taking readings of invisible signals like bacteria, radiation, or chemical balance. However, their use in this respect – to open up new forms of communication – has been limited. Fitbit, as already discussed, is an intensification of human-to-human computer interaction which reframes a subject of digital culture within a highly constrained and normative matrix of ideologies of self, fitness and productivity. What would it mean, however, to try to be faithful to the promise of new forms of communication such that a communicative back and forth could be established across different sensory worlds, and human and nonhuman agencies?

NEW MATERIALISMS

There are a number of art works that promise to take this direction, but many of them are, like the works cited above, one-directional. They take in biosensed data such as movement, sweat, electrical impulses, heart rate, chemical composition and use this as data to generate expression. This comes with a promise that the human sensorium will be opened up to a new horizon, but in most instances this isn't apparent in the form of expression taken, or, if it is, it is only apparent to an expert. For example, in the case of musical expression derived from heart rate or the sphincter muscle, a new musical range or new horizon of musical expression could only be discerned and judged in the same aesthetic register. Thus, while a professionally trained musician might judge this a new form of expression, an untrained ear would hear just another form of musical expression or collection of sounds. A more faithful approach, then, would involve an attempt to exploit the capacity of biosensors to bring new experiences to the sensible world.

A much smaller range of works have engaged with this question, some of them more conceptual than practical. One example is *M(y)crobes* by Stephani Bardin. This collaborative bioart project looks at biosensors to open up new understandings of how bacteria and humans interact. The following quotation from the project description by Bardin provides an idea of both its form and content:

> They have developed a wearable biosensor for the neck or wrist comprised of a disk of agar that is laid into a 3D printed bezel of recyclable ABS plastic that is exposed to the elements. Seeds will be implanted into the agar medium to track the microbial growth through the sprouting of these small plants. Thus, the wearer will see, in real time, the effects of their own biotic micro ecosystem in concert with the macro ecosystem of the environment.
>
> We share our regular world with billions of bacteria and fungi, but are for the most part unaware of how they shape our world, unless we get sick. The project aims to bring these organisms to the forefront by culturing them, allowing us to see how they directly impact a living being like a small plant, while also showcasing the variety of microbial cultures that we may encounter/host in our everyday wandering. These cultured microbes will actively affect the growth of seedlings placed in the agar-wearable, thereby allowing us to observe how these ubiquitous life forms actively affect life and growth by altering environments. (stefanibardin.net/mycrobes)

This project could be thought of as a more faithful attempt to open up the capacity of biosensors to bring new knowledge to the sensible world by making microbic life visible to people. It intersects with the contemporary interest in bacteria, microbes and the sequencing of the human microbiome (e.g. the Eden's Project's 2015 exhibition, *Invisible You: The Human Microbiome*).

The microbiome project was launched in 2008 and explores the possibility of genomic sequencing of the estimated 10,000 organisms that inhabit (cohabit) the human body. The kinds of organisms that are thought to make up the human microbiome are bacteria, yeasts, eukaryotes and viruses. This microbial turn in the sciences has resonance with the turn to 'green materialism' (Bennett 2009) in other disciplines such as philosophy, and with the material turn discussed in the introduction to this book. The connection across these fields is the desire to acknowledge and account for a much more radical assemblage of actors and agency in relation to both explanations of the world and interventions in it. In this case, the aim is to make visible the relational actions of microbes, humans and environment. Other versions of this direction could include works by Julie Freeman, who describes her practice as broadly 'translating nature' through works including *Lepidopteral* (2012), which uses kinetic sculpture to express environmental signals, or *The Lake* (2005), which uses signals from fish to generate an audio-visual interface. Other examples of environmentally disposed biosensory art include the Open Lab project *Oceanic Scales* (2015), developed by Gene Felice II and Jennifer Parker with their collaborators, which uses a combination of biosensors and biomimicry to look at pollution, human agency and marine life in the Monterey Bay.

In addition to reviewing art work, in the course of the research for this book I decided that playing with biosensors in a workshop format might be productive. The proposal was to invite a number of people to play with biosensor kits in relation to the question of how they might be used to vector communication between humans, nonhumans and environment. The ensuing events brought together a mix of people in the academy, in different roles and disciplines, and in different spaces. The first workshop combined a brief and a range of Arduino sensors. The different responses to the brief suggest an alternative account of what biosensors might be, although motorized cat ears might be biosensors' finest moment. The prototypes, ideas and projects that came out of the workshops were varied, but the process was also important in exploring some issues in relation to biosensors (for a more detailed account see O'Riordan et al. 2017).

In this workshop one of the first and most mundane issues was how to make the kit work. This also brought issues of uneven experience, access, capacity and differing expertise to the fore. In other words, are you used and confused by your technologies or can you intervene in them? There have been a number of responses to this question from coding literacy and education programmes, sponsored both by the state and industry; critical art collectives; hacking communities; activism; and critical engineering (Oliver et al. 2011). Technological innovation rather famously works to de-skill and demote specific groups of people and practices as much as it does to ensure progress (Bassett 2015).

Access and capacity have been discussed extensively in relation to digital culture, largely framed in the context of the digital divide and in terms of expertise, coding or computer literacy. Using a device like an Arduino, a widely used open-source basis for physical computing, falls somewhere in between reacting to an object like Fitbit and making it up yourself. Using Arduino kits means that you don't have to make things up from scratch, but they also bring constraints in that you can only work with what the kits provide. They still require some basic knowledge of coding, versions and code libraries, and the better the resources for coding the more things open up. In the workshop people had widely divergent experiences. We used a combination of sensors and e-health kits and this offered a wide range of fairly accessible sensing potential. There were still multiple issues within the workshop around making these work, identifying the right generations of code libraries (including different generations for different sensors), and rendering the data in a way that made sense.

The ideas that emerged followed directions not dissimilar to work elsewhere. One team came up with a blinking badge that expressed light in relation to heart rate (via pulse), where the pulse was taken from the earlobe and a wire joined the earlobe and a badge to transmit the signal. The badge itself was made out of mirrored perspex with an engraving of a flower. The idea behind the badge was to use sensors to engage with the notion of open-hearted communication. This was taken from one workshop member's recent experience at a permaculture-inspired activism camp where the language of open-hearted communica-tion had been used. The point of this project was to engage with the question of what open-hearted communication might mean and how it could be understood and engaged with. The prototype was imagined as something that everyone in a decision-making group could wear in order to be responsive to and considerate of nervousness and anxiety in communicative encounters. The idea of the mirrored badge was that you would see yourself in relation to others and thus could think about your own expressed light blinking and not only be motivated to surveil others.

Another response was a resistance to the call to play with biosensors as prototypes. For example, one team presented a manifesto as their contribution and read from a collective statement that riffed off the Critical Engineering Manifesto (Oliver et al. 2011). This move to resist the digital interface was in an inverse relationship to the expertise on the team. Indeed, one member of this team had been the most active in the workshop in figuring out coding solutions and navigating the issue of code libraries. This kind of resistance can also be seen in some of the examples above, including *M(y)crobes*, and in big data or quantified art work more generally. For example, Jacek Smolicki, whose work has been exhibited as part of the Quantified Self,[4] uses collages of news materials, portrait illustrations of other train passengers, and photos of found objects on daily walks. Collected by Smolicki under the collective title of 'on-going', this kind of work, although technically proficient and presented with high production values in web design, engages a post-digital aesthetic by eschewing data visualizations and drawing on illustration, collage and photography. *M(y)crobes* has resonance with a kind of post-digital making in its sculptural focus on the biosensor as installation, where the seeds and the petri dish are the interface rather than a representation of them.

MAKING MANIFEST

This chapter began with a reference to one manifesto and comes back to several others at its end. The act of responding to a maker brief with a manifesto helps to highlight some of the tensions between different framings of material and what constitutes material form or a material intervention. A manifesto has presence, performance and materiality; it has a material form in that it is written down, and it is declarative of presence and intention. It is read out, which performs presence and materiality, and beyond the moment of its reading it might cease to exist. Even as a declarative form it involves materiality: voice, breath, sound, hearing, echo, bodies that speak and hear, sound that is generated and travels. Those elements that could be said to be immaterial are the meaning of the sounds, the meaning making of the audience, the memory, trace and echo of the declaration passed into and through the bodies of audience and declaimers.

The etymology of the word manifesto is such that it draws together the meanings of a public declaration, and of things evident, obvious and made plain. However, the meaning is both the most immaterial and the most important element of the manifesto. What it makes plain is made so in an ephemeral moment of transition, conveyed through noise, but the substance of the manifesto is not that materialization but what is said. At the same time it is not only the trace but the event of the declamation that is material, as in germane, and this brings to the fore a

---

4 The Quantified Self refers to communities, events and people invested in quantification and involved in meet ups and other practices around this. See Lupton 2016 for further work on this.

tension between multiple states when all that is material fades into air, all that is ephemeral is material.

Another manifesto. Writing in 1999 about cyberfeminism, Caroline Bassett cautioned against finding utopia or the revolution in the realm of technology: 'This paper began life as a Manifesto against Manifestos. It ends as a call for the restitution of the idea of Utopia in cyberfeminism' (1999: 16). Her manifesto against manifestos is more specifically a critique of Sadie Plant's *Zeroes and Ones* (1998) as a manifesto. Bassett reads Plant's piece as a manifesto, alongside other declarative modes of cyberfeminism claiming that the revolution had already happened, and critiques these declarations for a kind of narrow tyranny. She argues that an engaged politics needs to reinstitute an idea of Utopia beyond the horizon, rather than celebrate technology. Her critique locates some modes of cyberfeminism in a similar terrain to that of the promissory futures of high-tech imaginaries because they both celebrate technology as liberation. This has resonance for thinking about biosensors, because in this moment engaged politics and technologies once again come together. The technological promise of biosensors is that through new forms of sensing beyond human capacity – or through changing dominant modes of human perception – we can see the world anew. This runs in parallel with the announcements in some materialist theory that the conditions of the present have an (already) radical capacity to make people see the world in new ways. However, the Fitbit vision of a population of self-monitoring, joined up, always on, productive, empowered and inspired young women is also a narrow and tyrannical framing of life. It isn't surprising that biosensing at its most commercial might also be at its most mundane. Thus, it makes sense to try and explore its avant-garde possibilities.

From motorized cat ears that read brain signals (Necomimi), to air-quality monitors that show us how polluted our environments are (e.g. Andrea Polli's *Particle Falls*), the experimental end of biosensing indicates more of a sense of utopian thinking beyond the horizon. Like much of digital art, biosensing work is often more interesting in conceptualization and process, and in terms of what it reaches towards, rather than the objects made. The radical promise of object orientated and speculative materialist theory is that the world beyond language will become communicable and that this will displace human centrality and bring about conditions through which we see the world beyond the narrow prism of capitalism or economic rationality. The use of biosensors to try to communicate signals beyond human perception, from our own sleep to the life of microbes, gestures towards that radical promise but also marks out its impossibility. All experiments with biosensors render the nonhuman through the human interface and tell us that there is much beyond that realm, but that it remains unknowable, elusive and unamenable to the register of systems of counting and measuring.

4

Smart Grids: Energy Futures, Carbon Capture and Geoengineering

Biosensors represent an approach to the world that breaks down everyday life into objects, generating devices that deal in data. Smart grids contrast with this vision of micro connections. They are unreal objects at a grand scale; as such they object orientate and they accelerate. They promise to join up power grids across regional, national and international borders in a new network offering real-time regulation of mixed energy sources. They promise instant reaction, response and regulation of energy flows and consumer behaviour. Multinationals like General Electric and National Grids such as the UK's promise to join up consumers, homes, businesses, energy suppliers and power grids in new and faster systems of control and redistribution. Smart grids allow a rebranding of energy multinationals as part of the solution to, rather than a cause of, global warming. They are an emerging technology that currently inheres in images and text, in policy documents, government statements, Super Bowl adverts, websites, posters, prospectuses and commercial briefing and strategy communications. However, like biosensors, they come down to small devices that generate objects from things that otherwise flow.

This rather wonderful description comes from the US Department of Energy's Office of Electricity Delivery and Reliability: 'Much in the way that a "smart" phone these days means a phone with a computer in it, smart grid means "computerizing" the electric utility grid.' Smart grids are about flexible, computerized power networks in which fossil fuels will become potentially less relevant and an increased use of renewables will come together with consumer awareness and agency. Smart grids promise that individuals and communities can control their own production and consumption of power in a series of networks making up a new kind of national (and international) power infrastructure. Outside of industry and policy domains, the most public-facing element of smart grids is the smart meter. Thus, although smart grids promise clean-energy futures, in the meantime they are also driving another mountain of post-consumer electronic waste in the shape of smart meters.

In Europe, smart grids are underpinned by the idea that energy production and consumption has to change. The vision of the smart grid is one of a flexible network that will enable a mixed renewables ecology to join up with fossil fuel sources. Key to this vision is the idea of controlled flow, so that a household with solar panels might produce, at times, more energy than needed, and would be able to pass that surplus back into the grid. At other times the production of renewables might enable the reduction or cessation of fossil fuel use. The grid, it is imagined, would connect power stations with consumers and producers of energy in a smart network where energy use is optimized through monitoring use at micro and macro levels, and real-time data analysis would be brought to bear on this.

The typical representation of a smart grid is a diagrammatic or textual account of a joined-up network of domestic and commercial customers linked to energy providers via distribution and control centres. Like the US Department of Energy, the European vision is also about comput-erizing the electricity grid, although the language of the digital is used instead: 'A smart grid is an electricity network that uses digital and other advanced technologies to monitor and manage the transport of electricity from all generation sources to meet the varying electricity demands of end-users' (IEA 2011).

This chapter traces these circulating visions, mapping different scales in an ecology of objects, and focuses on the role of public relations and advertising. In this case the public relations work of making smart grids is quite clearly part of their materialization. This communications work shifts smart grids from the abstract images of digital energy networks to an attachment to meters. The chapter explores the way that in the UK context smart meters have been promoted through relatively local attachments to nostalgia and heritage, to generate a vision of progress. At the same time visions of smart grids have become ever more expansive and reach for international connections at the expense of local ones.

The chapter also examines the use of the terms 'digital' and 'computerized' in these visions. Smart grids are a reminder that compu-tational networks are bound up in much older networks, and the chapter explores the possibility that smart grids also offer a way of rethinking network politics.

CIRCULATING VISIONS

The use of the term 'smart grid' has seen an exponential rise in the early twenty-first century. Although there were references to smart grids in relation to tax and to the Department of Energy in the US in the early 1960s, these were minimal. The term starts to appear in media sources around 1998 and has emerged in salience in the last 20 years. Currently a search on the term brings back over 35 million hits. Smart grids have a high media presence and references to them appear across multiple media forms and discursive fields. At the same time they have such uncertain ontology that they could be said to be non-existent (except perhaps as nodes in places such as Orkney). There are industries, trade presses and policies across multiple countries creating visions of smart grids.

Smart grids are a story told by energy policy bodies (e.g. OfGem (UK)), distribution network operators (e.g. UKPN), power companies (e.g. Enexis (NL); General Electric (US)), technology providers (e.g. Intel, Siemens), businesses investing in alternative technologies (e.g. Sainsbury's), governments, PR and advertising companies (e.g. BBDO), and the press. A network of technology press outlets and systems of industry recognition also contribute to a media culture of smart grids. In Europe the European Commission, the European Parliament and the Council of the EU have been significant authors of the smart grid story. In this promised future a macro energy grid, or network of networks, will combine multiple energy sources and measure need, or overload, and distribute accordingly, shutting off sources of power, or lights in empty rooms, and maximizing the resources of a mixed energy ecology.

In the UK the Department for Energy and Climate Change (DECC) was one of the primary policy actors in smart grids. However, the recently established (2015) National Infrastructure Commission is also an interim decision-making body in this area, and Smart Energy GB is another important UK actor. DECC published the *Smart Grid Vision and Routemap* in 2014, and a recent supplement to this is the *Smart Power* (2016) report published by the National Infrastructure Commission.

Visions of smart grids exist in these policy documents in which they are linked to statements about what will happen: 'Smart grids will enable new sources of energy and new forms of demand' (DECC 2014: 4). A future simple tense is often invoked to say that smart grids will be implemented and this will have predictable effects: 'Smart grids bring a range of benefits that will be felt in both the short and long term' (DECC

2014: 10). The use of these tenses invokes a definite time in the future, although DECC itself was shut down in 2016. Benefits are framed as definite, but the policy documents also use conditional and cautious language when it comes to the specifics of operational and economic questions, for example using 'could' instead of will: 'smart grids could lead to approximately £13bn of Gross Value Added between now and 2050; export earnings of £5bn to 2050 and jobs could be boosted by an average of 8,000 during the 2020s rising to 9,000 during the 2030s if sufficient investment is made' (DECC 2014: 10).

In DECC's *Smart Grid Vision and Routemap* 'will' is used 30 times while 'could' features 12 times. This indicates the rather unequal distribution of definite speculation in relation to cautious suggestion in policy. The Roadmap also references another report, *Smart Grid: A Race Worth Winning* (2014), which promises significant economic benefits to the UK. This report was prepared by Ernst and Young on behalf of SmartGrid GB (now Smart Energy GB), a consortium of companies including BT, British Gas, Scottish Power and EDF among others, and it estimates that moving to smart grids will cost £23 billion. This cost is overshadowed by the projection that existing systems will cost in the order of £140 billion over the same period. As well as this financial calculation, the report also projects significant gain to the economy through employment, exports and 'secondary industries' including electric vehicles and renewables. The Ernst and Young report is explicit that there is uncertainty: 'We recognise that there is inevitably an element of uncertainty over the individual findings quoted. However, we do not believe that these detract in any way from the key conclusion that the case for smart grid is compelling, robust across different scenarios and supported by international evidence' (Smart Grid GB 2012).

These visions and road maps all contain statements of uncertainty as to what the smart grid is while expressing great confidence in what it will do. Richard Tutton's (2011) analysis of forward-looking statements in US Securities and Exchange Committee (SEC) filings, in relation to biotechnology companies, argues that the promises of pessimism are also performative statements creating value, as much as are the hopeful promises. In this case there is no promise of failure of smart grid innovation but much promise of pessimism should the UK fail to take on leadership in this area.

Positive visions also appear in commercial advertising for smart grids in which there is less written text and much more imagery. The US multinational General Electric (GE) is a significant promoter of smart grids. In strong contrast with UK policy documents, their communications are orientated towards images with accompanying text. These are put together with caveats about the 'forward-looking statements' (General Electric 2009) that characterize commercial company filings (Tutton 2011). The GE 2009 briefing document *Smart Grids at Work* uses images in a ten-page publication and includes this standard text at the beginning:

> This document contains 'forward-looking statements' – that is, statements related to future, not past, events. In this context, forward-looking statements often address our expected future business and financial performance and financial condition, and often contain words such as 'expect,' 'anticipate,' 'intend,' 'plan,' 'believe,' 'seek,' 'see,' or 'will.' Forward-looking statements by their nature address matters that are, to different degrees, uncertain. (General Electric 2009: 10)

This attention to the form of communication and type of address would also be a useful precursor in policy documentation. However, policy documents are not regulated in the same way as company filings. It is interesting to note that there is no qualifying statement about the use of images in the GE materials (or elsewhere). There are regulatory frameworks and pro-formas for written text in economic filing documents in the US, and for documents with commercial and legal status in the UK. There is no equivalent for images, and questions about how to regulate the rhetorical force of images in other areas such as climate change remain unresolved (Mellor 2009). The rhetorical force of diagrams in communicating smart grids is central to making them an object.

General Electric has been one of the most significant producers of smart grid images. Their strap line is 'imagination at work'. They filled their 2009 Super Bowl advertising slot (the most prestigious and widely circulated advertising space in the US media) with a smart grid advert. They also ran an initiative called 'ecomagination', which includes an innovation competition for grid efficiency and renewables. In the 2009 briefing they provided the explanatory diagram of a smart grid reproduced in Figure 4.1.

SCALING DOWN: FROM SMART GRIDS TO SMART METERS

Smart grids often appear as images. As the regulatory context above indicates, images are a kind of object, unassailable and given. Unlike

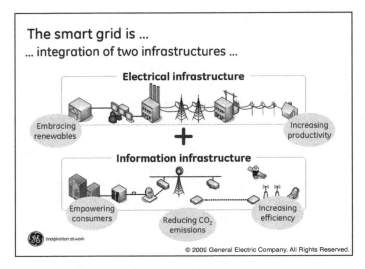

Figure 4.1 GE Image of a smart grid

written text, which is understood as made up and malleable, images, especially diagrammatic ones such as the example above, appear as objects. Also, unlike the smart phones in the US Department of Energy's allegory above, smart grids have not materialized beyond diagrammatic forms, although like other emerging technologies they are framed entirely as though they inevitably will. In the UK there is a potential smart grid of a kind, or a node at least, in Orkney, where the already heavy use of renewables is being managed in relation to the national grid and the Orkney power network through an active network manager (ANM). This is the only material instantiation of a smart grid node, beyond communications media such as policy documents and advertising. Much more widespread and in use, however, is the smart meter.

Smart meters are fairly widely distributed among the existing consumer base and have become synonymous with the smart grid. Although in fact they are in some ways antithetical to the smart grid aim of combating climate change because they instantiate a new industry in devices. Homes in many European countries are being equipped with them, and the UK government guidance to power companies is that they should aim to roll them out across the board by 2020. In the UK, 988,000 meters were installed by the end of 2012 (DECC 2012: 12). There has been concern about meters not working, their short shelf-life, and the lack of a central communications infrastructure, which means the meter is specific to supplier. In addition to the inbuilt obsolescence of these devices, which means the first and second generations are already aging before any smart infrastructure has been implemented, they also have to be replaced if you move supplier.

In the UK there has been a national campaign to roll out smart meters, coordinated through Smart Energy GB. Though funded by the energy companies, Smart Energy GB aims to stand independently of both them and the UK government. Its mission statement is as follows: 'Smart Energy GB is the national campaign for the smart meter rollout. It's our task to help everyone in Great Britain understand smart meters, the national rollout and how to use their new meters to get their gas and electricity under control' (Smart Energy GB 2015: 4). Smart Energy GB is about communicating smart grids in order to effect consumer change to enable this roll out. Accordingly, much of their operating budget is spent on communications, and they are a client of the prominent UK advertising agency AMV, part of the international BBDO agency, owned by the OMNICOM group.

In 2015 AMV ran a campaign which linked a cinema theatrical release, YouTube videos, the Smart Energy GB website, multiple press releases and social media activity. The campaign was based around the animated characters Gaz and Leccy (gas and electricity respectively), who were represented through a short animated film as embodying out of control energy use, in order to position smart meters as empowering consumers to retake control. The animated characters chased each other round a house at night time, turning on appliances, shorting out circuits and causing sparks and small fires and explosions. The representation of gas and electricity as animated characters also invoked the idea that these forces are themselves alive, and late nineteenth and early twentieth century tropes of electricity as a kind of magical force or animated sprite played out in the animation. There were also references to horror tropes, notably *Poltergeist* (1982), in which the television screen and electrical forces are part of the manifestation of horror. The animation was overlaid with the 1970s UK glam rock track, 'Ballroom Blitz' by Sweet. This way of representing electricity and gas invoked a nostalgic aesthetic, positioning the way that power works now as old-fashioned, out of control, and belonging to the past. Smart meters were offered as way of banishing older unruly forces and controlling them with new devices.

Smart Energy GB and AMV's media intervention provides a different way of representing smart meters, which according to their own research only around 20 per cent of the population understand (2015 annual report). The same data sources estimate a 12 per cent propensity for take up, which apparently jumps to 35 per cent once people are exposed to the Gaz and Leccy campaign. The campaign operates in a context where there is a vacuum in the representation of smart grids themselves. Devine-Wright et al. (2010) note that media representations of smart grids in the UK are overwhelmingly attached to images of male politicians and to wind turbines. While media coverage of smart grids has grown exponentially since their study of the 2009 UK media, representations are still either abstract and connected to diagrams, personalized in relation to specific politicians, or connected to images of renewables such as the wind turbine.

In this space there is a relative vacuum in terms of communicating smart grids. The AMV campaign introduced the Gaz and Leccy animations and focused on the meter. Related to the AMV campaign, and also underwritten by Smart Energy GB, is a focus on the disappearance of the traditional meter. Two art projects, one poetry and one classical music, have also become part of this site of public engagement with smart meters. These projects represent 'old' meters as part of a nostalgic history of Britishness. Thus, smart grids approach the question of what they are through the making of an object, and once the object is made, like other objects it becomes a given. Reinforcing this sense of inevitability is the making of a history around meters. Unlike genomes, where the making of sequences as objects requires an erasure of histories, the making of smart meters as the object of smart grids entails history making.

Placing these meters in the past and aligning them with twentieth-century cultural forms and heritage signals the new generation of smart meters without having to focus on them. The Royal Philharmonic Orchestra performed *Requiem for Meters* (2015) using instruments made of old gas and electricity meters. All of these were large-scale meters, quite unlike those in any new build, associated with an older image of twentieth-century history. The recital and making of the piece were also released as video materials. It was recorded at Abbey Road studios, synonymous with The Beatles and reinforcing the post-war nostalgia feel constructed around 'old' meters.

The British poet laureate, Carol Ann Duffy, was also asked to create a poem about the roll out of smart meters. The commission was to focus on celebrating the passing of an era. In a press release in April 2016 Duffy was quoted as saying:

Gas and electricity meters have been a fixture under stairs and in cupboards for more than a hundred years so it felt fitting to preserve their place in household history with a poem. It is definitely one of my most unusual projects, but hopefully I'm able to produce a piece that captures the last whirs of these spinning machines before they make way for their digital counterparts. (Smart Energy GB 2016)

Sacha Deshmukh, Chief Executive of Smart Energy GB, also commented: 'There is a great British tradition of marking national moments with poetry.' The press release included photographs of Duffy looking at old meters, and mentioned the *Requiem for Meters*, constructing meters as part of a British cultural tradition. The imagery associated with smart meters in both cases was constructed through a representation of a particular kind of meter as national heritage. This contrast between the old objects and their 'digital counterparts' makes traditional meters emblematic of a vision of an analogue past, and smart meters aligned to new digital present and future. This curious temporality in which nostalgia reconstructs the digital as a new (again) technology aligns smart grids with the network society of the 1990s (Castells 1996), as well as the big data imaginaries of the moment.

Most positive representations of the smart meter have focused on consumer control and projected changes to domestic energy use. This quote is from research into households issued with smart meters:

Habits in the house have changed, we don't leave computer screens on now, we turn them off instead of leaving them on screen saver, turn them off at night, laptops they turn them off at the plug, don't leave them on charge ... and when kids took their clothes off at night, chuck them in washing machine, half load, but now when you see how much a washing machine can affect the level I wait till it's a full load, that makes a big difference. (DECC 2012: 3)

Smart meters are thus framed in terms of efficiency and productivity and linked to changes in individual behaviour, enabling data collection about everyday life. This framing is very similar to that of the biosensors examined in the previous chapter. They generate data about behaviour to construct knowledgeable consumers and encourage positive energy choices.

However, smart meters are not really smart in the broader sense of the vision promised. They measure energy use but they don't connect to a joined up system or enable direct changes in energy flow because there isn't a smart grid in which they can connect. They are isolated devices which bring measuring closer to the consumer. Electricity and gas companies already measure individual and collective use, peak use times, down times and moments of disconnection or drainage of power. The direct address to the consumer is an important change, and the visibility of counting at the individual level creates new kinds of knowledge. Smart meters can prompt consumers to think about energy use and create a new perception of it through real-time monitoring. It is one thing to know that boiling an electric kettle takes a lot of power and is expensive over time, it is another thing to see your electricity use spike in real time as you turn it on. However, like patterns of use associated with Fitbit, although this has an initial wow factor, there is much evidence to suggest that once the novelty wears off, long-term habits don't change that much (Buchanan et al. 2015).

Figure 4.2 British Gas Smart Meter alongside a 'traditional meter' with Carol Anne Duffy

It is possible that smart meters might help people to be more conscious of energy use and be impelled to action to reduce consumption. However, even if they do so, they also undo the smart grid vision because the meters have significant contradictory features in terms of delivering a less wasteful energy future. Firstly, many are already old and not smart; first generation models have already been replaced by new generations of devices. Secondly, the new smart meter production industry is contributing to the exponential growth in personal devices that populate our worlds.

Rather than helping people use less energy, smart meters add to computerized device production with all the increases in energy consumption, problematic materials and uncertain after-lives that we already see with mobile phones, tablets and computers. Like other individual consumer devices, they very quickly join the vast and proliferating new forms of electronic waste. As Jennifer Gabrys observes: 'The proliferation of computational things within the Internet of Things reads as an itemized list of electronic waste in the making' (2016: 182). In the UK, the 2015 generation of smart meters are supplier-dependent, which means that if people change energy suppliers the meters don't work, which then generates demand for another round of meters. Thus, although smart meters have become the symbol of the smart grid in relation to national suppliers and consumers in the UK, their capacity to materialize enormous amounts of post-consumer toxic waste, and to deplete resources further, is antithetical to the avowed aims of the smart grid. Smart grids promise ultimately to deliver a more efficient, more renewable way of managing energy and resources, but in the making of an object smart meters materialize a new generation of devices that will become, very quickly, rubbish.

SCALING UP: FROM SMART GRIDS TO ENERGY FUTURES

Smart grids have materialized in the questionable form of smart meters and in the UK through a specific imaginary of national culture. However, the smart grid imaginary also scales up to visions about grids, networks, nations, the planet, the atmosphere and outer space. Smart grids are an imaginary in which a vision of planetary control is possible. Like the Internet of Things, they are open to links with other networks and a potentially global power network. For example, in the UK there are ideas about linking the national electricity grid to Iceland's geothermal sources through an undersea power line.

These visions demonstrate the same patterns seen in other areas of promissory technoscience: they offer world-changing technologies which also enable business as usual. They reproduce the god trick of the view from nowhere, a view that promises clean diagrammatic control over an impossibly chaotic and dirty world. Smart grids are a visualization rather than a vision. They can't be seen but they promise a visual imaginary of networked devices and an Internet of Things. One scaled-up version is that of multiple grids and super-grids in which localized smart grids link to others, becoming network nodes in an indefinite network imaginary. However, there are also other indefinite imaginaries in which different kinds of practices are imagined. I explore two examples here to draw out some of the politics of these unreal objects: Desertec and carbon capture.

Desertec is a vision for solar energy in which deserts become the new energy producers of the world. In particular the northern deserts of the African continent become the power sources for Europe: 'In our perspective the key to decarbonize the world and fight global warming is the desert – in only six hours deserts receive as much energy from the sun as humankind consumes in a whole year' (desertec.org). Desertec is a specific project, although still an imagined one, in contrast to the more generic imaginary of smart grids and carbon capture. There is a Desertec consortium and the project has been discussed extensively in the energy policy literature (e.g. Lilliestam and Ellenbeck 2011; Samus et al. 2013). A group of shareholders set up a company in 2003 and considered a proposal for operations in the order of 400 billion euro. For a while Desertec gathered together a coalition of actors, and it has had some salience in European energy politics (Rothe 2014), but as of 2014 only three of the original 50 shareholders were left (Reuters 2014).

Desertec is interesting to think about in relation to Bruno Latour's analysis of the imagined urban transport network Aramis, which he subtitled 'the love of technology' (Latour 1996). In the story of Aramis enchantment is central, and Latour helps to dispel the idea that technologies are somehow divorced from emotion. Latour's analysis was an indication in science studies that technologies can be about failure and rather obviously generate affective attachments. The whole terrain of unreal objects can be thought of in the genre of a romantic drama, casting technologies as love objects. As with Aramis, this is a dedicated and mystical worship as well as romantic love, where technology is cast as both redeemer and seducer. Aramis was an imagined network which saw investment, trials and experimentation over several decades. The Desertec project also has other antecedents in desire and utopian thinking, such as an earlier twentieth-century project called Atlantropa that promised to build a hydroelectric dam across the Gibraltar straits in order to increase the land size and to power the Mediterranean region (Ley 1964).

The idea of Atlantropa was also part of a dream of domination in which Europe and the US would remain and grow as major world powers. The continent of Africa largely figures as resource for Europe in this perspective. The project materialized in a series of publications between 1929 and 1948 by Herman Sörgel, a German architect who authored the idea. It also travelled through the discussions which circulated around this, including illustrations in the press, and documentaries. It even entered into science fiction in Philip K. Dick's 1962 novel *The Man in the High Castle*, about an alternate history in which Germany wins the Second World War, and featuring the draining of the Mediterranean and the destruction of Africa as part of this history.

Like Atlantropa and Aramis, Desertec can be thought of in terms of an engineer's dreams (Ley 1964). Unlike Atlantropa, it imagines alliances rather than domination. It also constructs a utopia of political alliances throughout the Mediterranean area, Europe, North Africa and the Middle East. Desertec's vision is one of resources flowing between these areas in a future in which the region is no longer divided by the existing political conditions of animosity and violence. It calls for a collective political will to enrich North Africa and clean up Europe in relation to energy futures. Africa is still framed as a resource for Europe, but one in which North African nations have agency: desert power.

Desertec materializes primarily in textual form, in discussions and alliances, but it also has more significant material nodes. For example, large solar installations are already a feature of the Mediterranean area and parts of North Africa. Two of the largest concentrations of solar panels are the Nzema Project in Ghana and the Ouarzazate solar power plant in Morocco. These are not yet operational but Nzema was projected to be so by 2017. Both are referenced in the Desertec plans. Ouarzazate aims to export power in future, but both look more like local power projects at present. Delf Rothe has pointed to the ways in which Desertec is a media project:

> it is not easy to classify the Desertec concept as it transcends the established categories of political or economic actors. It is, at the same time, a business consortium, a lobby group, a civil-society initiative and simply a discursive vision ... one can best describe it as a discourse coalition that has developed around a set of storylines bound together by the highly charged signifier Desertec. (2014: 7)

Unlike Aramis, which materialized in some forms (experimental train cars, airport trains), Desertec remains what Rothe refers to as 'a discourse coalition'. Although parts of the technological whole are viable – for example the solar panels – the visionary aspects are not about techno-logical breakthroughs as such (there are many viability questions around energy storage and dissemination), but about new political formations, political agency, decision-making and communication, and in this sense they are not really simple. They are more politically visionary than the smart grid projections of General Electric because they are much more about people and politics rather than technological innovation.

In the case of Aramis numerous trials of new technological ideas (e.g. carriages that were linked but could also break off from the train to take different routes without requiring decoupling) were carried out but posed problems that at the time seemed to be technically insurmountable (Latour 1996). In the case of Desertec the innovations that would be required to realize the project are new political and economic mechanisms to enable cooperation across a broad collective of nations. This collective imaginary differs from that of smart grids in the UK, which although framed as a collective effort to address climate change, is enacted through the production and consumption of individually owned devices.

OUT OF SIGHT, OUT OF MIND

Carbon capture and carbon credits are elements of the smart grid energy imaginary played out in policy and energy businesses in the UK and globally. The same key actors promoting smart grids are also promoting and investing in carbon capture technologies. Carbon capture promises something like business as usual in that we can continue to generate CO_2 emissions at the same rate, but ameliorate climate change by capturing and storing the gases somewhere outside of the climate.

One example is the proposal to bury emitted gas under the floor of the North Sea (DECC 2012). The idea is that the chambers created by the extraction of fossil fuels become the spaces of storage for carbon gas. This echoes the handling of nuclear waste in its out of sight, out of mind mode. Another proposal is to turn gas into stone, and experiments with solidifying gases in this way have been carried out in Iceland (Matter et al. 2016). In this process the gases are dissolved in water which is then pumped through basalt rock, creating a mineralization of the carbon similar to limestone formations. This is a beautifully gothic imaginary of transfixing the monster in stone, and has multiple references including the monsters turning into stone with the appearance of daylight.

Carbon capture is one of the many proposals on the table for meeting targets to reduce carbon emissions in order to address climate change.

Carbon capture refers to a very general idea but also to specific plans, and in the UK context such projects have so far failed. A recent plan was for the Drax energy company to capture the carbon emitted from the power station of the same name in North Yorkshire and put it into the empty reservoirs created by drilling in the North Sea. However, Drax pulled out in autumn 2015 after allegations that the UK government was failing to support the development of carbon capture in economic terms. The now defunct DECC continued to promise that carbon capture was part of its strategy.

These examples all demand technical innovation in relation to energy production, storage and dissemination. They are also political ideas and would require innovations in politics, economics, consumption and communication in order to succeed. However, the focus on those aspects framed as technical take centre stage in energy policy, even though the political questions about how to work collectively on these issues are more pressing.

SMART GRIDS AND GLOBAL WARMING

Thinking about global warming through the lens of unreal objects raises the question of what kind of thing it is. Climate change has a very high media presence and an uncertain ontology. There is however, a scientific consensus that anthropogenic global warming is real, and has very material effects, such as melting ice, rising sea levels, changing weather, flooding and drought. It has become relevant in terms of famine, war, displacement and refugee crises, and in many places it is experienced and felt. For example, in Iceland overall warming has accelerated since 1975 and the consequent lifting of the ice cover has enabled the land mass to rise and volcanic activity to increase. Plant and animal life has also been affected, with changing fish populations and increased arable production.

The war in Syria, although focused on regime change, was also triggered during a long period of drought which displaced the rural population into urban areas and was a factor in the civil war (Gleick 2014). The war has caused major devastation and casualties in Syria, where half of the population has been displaced and over 250,000 people have been killed as of 2014 (Data Team 2015). This has precipitated a refugee crisis in the region as well as in Europe. At the same time small islands and coastal areas in the Pacific (Ferris et al. 2011), the Caribbean, the Gulf of Mexico and Alaska have seen a loss of surface area that threatens communities and is forcing displacement. The displacement of people is a global issue and will likely increase. The UN refugee agency reported that the displacement of peoples was at an all-time high, with over 65 million in 2015, half of whom are children (UNHCR 2015). These are realities and materialities.

Environmentalists and climate change scientists have been accused of something like a Cassandra complex because of their warnings about the problems of global warming and climate change. Like the mythical figure, they have been framed as propagating fraudulent, sensationalist and scaremongering visions. In contrast, the energy futures promised in the smart grid industries and policy arenas are taken as 'forward-looking statements', offering predictive analysis and invoking the future in the present through a framing of factual communication. The smart grid industry is in many ways a reframing of the energy industry as poised to save populations from climate change. This entails an incorporation of climate change threats into the discourse of the industry, mainstreaming climate change, in order to position the industry as an actor with the capacity to ameliorate the threat. At the same time this relocates technology as the solution, as well as the problem.

This nexus of problem-solution framing has a resonance with Bernard Stiegler's (2010) discussion of the Pharmakon. Stiegler positions technology as both disease and cure and warns against extreme responses to both sides. His analysis shares the technocentricism of technology and innovation discourses. It is tempting to think Cassandra and the Pharmakon together in this case. In both classical and psychological treatments, Cassandra is a woman gifted with prophecy but never believed. In psychological discourse she has been figured as internalizing this dilemma, connecting it to discourses of hysteria and gendered pathology. In classical myth her prophecy is accurate and she is tortured by the curse that she will never be believed. The Pharmakon on the other hand is figured as a story of technological evolution, in which the disease of technology will be cured by the same. Warnings about climate change, however, demand different kinds of solutions, such as social and political change and reductions in technology use. The suggestion that we should use less technology almost amounts to blasphemy in current technoscientific cultures. Cassandra can only be reformulated as cure, or refigured as Pharmakon, through the promise of geoengineering and biotechnological solutions to climate change. For example, de-extinction or carbon capture.

The Cassandra complex in psychology is also used to explain how those who warn against things to come are also then blamed for causing them. This resonates with Sarah Ahmed's (2010) discussion of the feminist killjoy who is not just the one who warns against being carefree, but who is also blamed for causing the burden of care (the killing of joy). Instead of the figure of femininity, Stiegler's Pharmakon figures technology as prophetic, promising both utopia and disaster, but also alleviating the worst of its own consequences. This poison-cure dichotomy is opposed to the figure of Cassandra and the feminist who prophecies and is also blamed. In the model of the Pharmakon, in the scientism of culture (Welsh and Wynne 2013) and the ideology of technology, technoscience brings disaster but is not blamed – it is always able to be rearticulated in relation to a future promise and cure. These are the conditions that enable the endless recuperation of technology in future-orientated fictions, utopian visions and cures for what ails us, at the same time that its forms are also those which create the conditions we live in.

One example of these conditions is the concept that there is an outside of the climate. Putting greenhouse gases under the sea-bed is part of a logic of fragmentation, separation and externalization (perhaps a digital logic). Like the earlier and utterly disastrous ideas that toxic waste could be buried or built over, or that radioactive waste could be cased in concrete, the idea that things can be isolated from the world within the planetary ecosystem betrays a logic of disassociation and fragmentation. In common with the use of dissociation in psychological discourse this could be seen as a coping mechanism triggered by stress or trauma, and as with the psychological simile, if not taken seriously as a problem it is likely to end badly.

What kind of realities are made in the construction of these unreal objects? Devices, discursive coalitions which construct their objects as real but obscure their materialities, industry sectors, policies, reports, prospectuses, books, news. On the one hand we are asked to attend to materials, but on the other materials become more informational or externalized. If we are asked to attend to things which amount to high media presence and ontological uncertainty, how then does it become possible to act in and with the world? Political capture of the generative capacity of mediation has enabled the production of such a multiplicity of discourses and discursive coalitions, and thrown into uncertainty what it means to have material effects.

In such conditions of over-representation and uncertainty, what are the possibilities for gathering around matters of concern and making a meaningful intervention? New technologies reproduce the conservative ideologies in which they are entangled, and in this case they immobilize change, or enable action which maintains the status quo. Carbon capture and the making of more devices reinforce the idea that consumption is agency (and is political agency) and that fossil fuel industries will continue, just more cleanly. However, is it possible to adjudicate between the instability of putting carbon back under the sea and the instability of building solar panels in the desert? Or – whose media realities get to materialize?

Communication about climate change has proved vexed. In the late twentieth century the representation of the growing global warming consensus was undermined by multiple interested parties. However, there were also issues in the mode of news reporting itself. For example, the journalistic adherence to showing both sides of a story, or the imperative for objectivity which demands that journalists look for dissenting voices, contributed to uncertainty about the breadth of the consensus. This patterning gave a favourable bias to minority dissent and public relations materials from energy companies in news reporting, particularly in the US (Boykoff and Boykoff 2004). This 'balance as bias' contributed to the construction of uncertainty about global warming.

However, other issues in the modes of communication available have also made this area difficult. Julie Doyle's analysis of climate change communication found that the dependence on visual evidence, particularly the photograph, posed challenges. It meant that understandings of climate change were often very abstract and thus made distant (temporally, spatially and experientially). Where photos were used as visual evidence, for example to show a before and after framing of glacier reduction, they created a visual image of change that indicated it was too late to do anything. 'The moment climate change can be photographed is (…) too late for preventative action' (Doyle 2007: 146). Once symptoms are photographed, such as melted glaciers, this frames the situation as too far gone. This further disempowers audiences, and destabilizes the capacity for climate change publics, because it seems as though nothing can be done. Doyle has argued that much more attention to the unseen aspects of climate change is necessary, and also that a broad coalition of actors needs to be involved in communication, and a focus on agential and positive communication must be a part of this.

This raises the question of why some speculative futures gain political traction as opposed to others. The forward-looking statements of the energy companies appear to inspire politicians, investors and engineers, despite their fictional status and, in knowledge assessment terms, their lack of robust knowledge (Von Schomberg and Funtowicz 2007). How then might other speculative futures gain traction, such as ones in which humans could act collectively to change the conditions of anthropogenic global warming, without exacerbating those conditions through unstable geoengineering propositions such as carbon capture, or the production of more of the technological interventions that created these conditions in the first place, like smart meters.

NETWORK POLITICS AND POWER: THE GRID AND THE NETWORK

In the promotional materials and visions for smart grids the emphasis is on adding computation to electricity. However, media history demonstrates that communication technologies are a process of making electricity visible (Batchen 2006). Electric media and electronic media are terms in media archaeology that frame communications as media form and infrastructure as much as message or content. McLuhan's media theory is concerned with electronic media and the sense of change that he felt came with both the form and the content of electrical media. In McLuhan's analysis the light bulb is a medium as much as television. His thesis was that we live in an age of electric media which extends the human: 'we have extended our central nervous system itself in a global embrace' (McLuhan 1964: 4).

The idea of the network society is based on the same premise that electronic media represent a different kind of era. Castells argues that electricity and microelectronics have enabled a social shift such that vertical power structures (institutions) have become less powerful and the power of networks is the order of the social (Castells and Cardoso 2005). Castells, whose work promotes the idiom of the network society, writes: 'Digital communication networks are the backbone of the network society, as power networks (meaning energy networks) were the infrastructure on which the industrial society was built' (Castells and Cardoso 2005: 4). His extended work *The Network Society* looks at the ways in which the new networked conditions of the later twentieth century and onwards create new forms of dominion, in particular that of the space of flows over the space of places. Other scholars including Hardt, Negri, Terranova and Lovink all deal with the idiom of the network in different ways, but what they have in common is the use of this language to refer to a communications network that is also a mode of socio-technical political power. Terranova, for example, writes:

> Here I take the Internet to be not simply a specific medium but a kind of active implementation of a design technique able to deal with the openness of systems. The design of the Internet (and its technical protocols) prefigured the constitution of a neo-imperial electronic space, whose main feature is an openness which is also a constitutive tendency to expansion. (2004: 3)

Castells examines activist networks and the mobilization of dissent in his later work, *Networks of Outrage and Hope* (2012), which, as the title conveys, foregrounds the space of activism and social change rather than that of global dominion. Other theorists see the network as an extension of the means of oppression of one class by another (e.g. Dyer-Witheford 2015). Both Jodi Dean's (2005) formulation of communicative capitalism and Nick Dyer-Witheford's of cybernetic capitalism cast networked communications as forms of circulating capital. In Dean's analysis the circulation of communicative forms depoliticizes communication itself. In Dyer-Witheford's account the power of the network goes beyond the accumulation and intent of human agency so that the system itself becomes an agent and node of power. In these modes of analysis humans become little more than a reproductive function of capital and, by extension, of computers.

The network of networks, whether the technical infrastructure of communications or the mode of politics, is a nonhuman agent of global proportions. Although the network has been the emblem of contemporary politics and communication technologies alike, the grid has also operated in relation to this core term in different ways. In *Network Culture*, Terranova's use of the term 'grid' implies something more like the vertical systems of power of the pre-network society. She describes the grid as a modernist figure, a symbol of the power of the rational mind over the chaos of topography, of the exercise of rationality and order (2004: 46). The contrast between the dynamic and open features of the network as opposed to the grid are clear in this quotation:

if the Internet were nothing but an electronic grid or database where all locations lie flat and movement is mainly that of vectors of fixed length but variable position linking distant locations to a few centres – where would the potential for struggle and change, becoming and transformation come from? In the case of the Internet, for example, where would its dynamism come from? How can we reconcile the grid-like structure of electronic space with the dynamic features of the Internet, with the movements of information? (2004: 49)

This contrasts to the use of the term grid in Castells' work. Here the grid is the new more flexible mode still to come, which (in 2005) he imagines will manage wireless networks. Contributions to his collection on policy draw on the language and promise of grid computing and wireless architecture, to articulate a future grid that will be more flexible and open: 'scaling-up the current patchwork of community access points into a larger grid that provides a true connectivity alternative for those [with] limited technical expertise and for local institutions with more complex service demands' (Bar and Galperin 2005: 278).

As a forward-looking term, promising greater processing power and more powerful distributed computation, the idea of the grid has been in circulation in computer science since the mid 1990s. It is currently associated with bioinformatics and big science projects like CERN. There has been considerable investment and development in European grid-computing infrastructure, which is also ongoing. Grid computing, as authored by Ian Foster (2000) for example, is targeted towards specific goals and aims to create protocols that allow pooling of resources so that scientific research can be spread across a network. Driving these ideas is the spectre of big data, and the promise of big science – specifically that of CERN and the life sciences.

Grids and clouds have come into play at the same time; one of the main distinctions drawn between them in IT discourses is that grids are open source and task orientated (e.g. the Large Hadron Collider), while clouds are proprietary and ongoing (Facebook, Amazon Cloud). Foster et al. (2008) compared cloud and grid computing in an article which aimed to give insights into both. The article also operates to reinstate Foster as the architect of grid computing and to emphasize the development of grids in an academic and science research environment as contrasted with the commercial trajectory of clouds. In this they note:

In the mid 1990s, the term Grid was coined to describe technologies that would allow consumers to obtain computing power on demand. Ian Foster and others posited that by standardizing the protocols used to request computing power, we could spur the creation of a Computing Grid, analogous in form and utility to the electric power grid. (Foster et al. 2008).

In the visions of both grid and cloud, computing power is referred to as being like electric power, or like utility grids. The point of the analogy is that computing power will be available on demand through shared protocols that standardize a system, allowing access to as much computing power as is required for any given project. This represents the aim of both grid and cloud as being the same, enabling the connection of any device to the electricity network, pretty much anywhere, and using as much power as required on demand. The power here is computational processing power instead of electricity. However, the picture is a bit more complex than in the case of electricity, which is the same across multiple uses. Software, data, information, code, programming, files, processing and so on form a much more heterogeneous set than that of electricity. What also disappears in this vision is that computing is a media form and that the history of computing also involves the idea of 'making electricity visible'. In this analogy electronic media are reduced to electricity, yet computation will be applied to electricity networks in the smart grid vision.

Smart grids envision both a new computerized electricity network and a computer network that operates like electricity. Both bring together computing power and electrical power in a vision of joined-up flexible, unlimited systems. However, computing power demands a huge amount of electrical power. As numerous studies have shown, the digital economy has a very significant energy footprint. In the UK government figures for domestic electricity consumption are broken down into six categories: consumer electronics; wet appliances; cold appliances; light; cooking; home computing (DECC 2014). Consumer electronics and home computing are counted as separate categories, even though they are increasingly hard to separate out; charging a phone, watching Netflix, or playing a networked game comes under consumer electronics, for example. Consumer electronics are already the biggest factor in domestic electricity consumption; taken together with home computing, the two categories account for the consumption of two thirds more electricity than any other category.

ALTERNATIVE IMAGINARIES AND ACCELERATION

Liz Jensen's novel *The Rapture* (2009) imagines a future in which drilling for methane gas will trigger an apocalyptic moment in an already post-apocalyptic world. This mode of speculative storytelling is a kind of accelerationism – speed things up and they will crash into a new reality – but with very different politics to the version of accelerationism that appears in object orientated materialisms.

Accelerationism in its relation to digital media theory is a discourse or a series of propositions, positions and manifestos. It has been influential in opening up debate about technology and culture. Associated, in its left-orientated formations, with Benjamin Noys, Alex Williams and Nick Srnicek (2013), it is associated with the idea that speeding up – rather than putting brakes on – processes of capital accumulation and technological change will lead to some kind of revolution, or a radical transformation of the world and what it means to be human. In some versions this is because the trajectory of endless progress, innovation and growth is unsustainable and will lead to an inevitable collapse. In this story, capitalism hastened will simply eat itself. Another version of accelerationism proposes that intensities of technological change will produce radical ontological shifts, such as the singularity, or new unimaginable forms beyond the horizon of possibility. In this sense, the accelerationist manifesto and other variations of accelerationism offer fatalistic, techno-utopian, de-politicized visions of an object world given to us in which we can only respond.

Much of the debate around accelerationism has hinged on the question of the capacity of accelerationism to be political at all (Power 2015). For example, Nina Power argues that accelerationism could only become political if it considered labour, subjects, violence and the practical work of political intervention. Other trajectories of debate that could be traced through accelerationism include deceleration and slow movements – slow politics, food, time, work – and xenofeminism. The latter is an explicitly feminist response to the accelerationist manifesto which cuts right into the lack of subjects in accelerationism: 'Technoscientific innovation must be linked to a collective theoretical and political thinking in which women, queers, and the gender non-conforming play an unparalleled role' (Cubonicks 2015). This critique of the erasure of subjects in accelerationism help to show its connections with object orientations and forms of speculative materialism.

Even sympathetic discussions of accelerationism acknowledge its 'somewhat unsavoury macho tone' (Brennan 2013). Cubonicks (2015) and Power (2016) call out its anti-feminism, constituted through tone, citation and epistemology. The complexity of subjectivity, identity and difference get brushed aside by the theory of grand theory, and by the technocentric, elitist scale. Accelerationism, like other theories of technology and culture, takes on everything from the micro level of the nanoscale to the intergalactic and universal. It is epic. For all its political promise, lives are not lived on those scales and any political intervention has to be rooted in the political realities of everyday life and its mediated materialities. Accelerationism and object orientation seem too far in tune with the technological objects they accept as givens to be able to address them as political sites of struggle.

Jensen's novel extrapolates from debates in industry which could be read as modes of accelerationism. For example, one argument for using methane gas is that global warming is already destabilizing the conditions in which it is secured in deposits, and this is already enabling its release into the atmosphere. In other words, global warming could accelerate more sudden global warming via methane leakage (Parmentier et al. 2015). Methane is a greenhouse gas, like carbon dioxide, and methane leakage is a risk factor in shale gas mining (fracking). The instability of methane gas deposits indicates the impossibility of securing carbon capture. Events such as seismic activity, temperature and geological and physical changes to those areas that have already been destabilized through drilling and mining activity will also enable the release of carbon dioxide 'captured' in such reservoirs.

Carbon capture and methane fracking are control narratives, in tune with object accelerations and a Victorian mode of industrialization. Today, when the instability of the grounds of existence and the unpredictability of human agency has been opened up, they seem like misplaced fictions. *The Rapture* contrasts these with a well-placed fiction, in which a vision of an utterly unpredictable and unstable reality overwhelms human agency.

With its sense of making things right by putting things back, carbon capture is an object orientated reparative narrative. It offers the promise of restoring things to a different temporal point, before the release of the carbon from fossil fuel already burnt. This reparative dimension, or sense of putting things back both temporally and spatially, connects it to the de-extinction narratives explored in the next chapter. Those narratives use fictions about restoring the past to promote biotechnological interventions like cloning. Carbon capture offers a similarly fictional reparative turn and evokes a similar atmosphere of nostalgia. Taking seriously a range of fictions and taking responsibility for them, from science fictions to share prospectuses, and including activist and artist imaginings of climate change, is one strategy for intervention. Looking at the role of different modes of representation also opens up other possibilities. In Doyle's analysis of the visual culture of climate change the emphasis is on visual evidence and a history of visual knowledge as producing truth. Currently there is an even stronger orientation towards data graphs, models and data visualizations of multiple kinds.

ORKNEY ALTERNATIVES

Another way of thinking about alternatives is to look at one of the few places where living with renewables is part of everyday experience. Scottish and Southern Energy (SSE) have trialled a smart-grid-like set up in Orkney where over 100 per cent of the islands' energy needs are already met by renewables. This area was a late recipient of the national grid: there is only one connecting power line and unstable connections have always been an issue, with back-up generators and alternative sources of power the norm.

The SSE trials are only the latest in a long history of Orkney alternatives. The use of renewables in an archipelago surrounded by sea and with the windiest weather in the UK makes sense, and has many historical precedents. The islands host the longest running, largest and most productive wind turbine in the UK. This single turbine generates enough power to support 1,400 households (Munro 2015), approximately 14 per cent of the estimated 10,000 Orkney households. SSE, in partnership with the University of Strathclyde, have also set up an active network management system (ANM) of the kind required for any smart grid future.

One of the challenges with the smart grid idea is that a complex network management system that monitors flow, use and a mix of power sources would be a necessary innovation. Orkney is often lauded as leading the way with renewables (wind power), sustainable power use (electric cars), and now smart grid management. However, innovative as these real conditions are, there are also multiple ways in which this is an exceptional situation. A community which is very differently located in relation to the national UK imaginary of centre and periphery, in an area which has never been fully integrated into an always-on unlimited power grid culture, is in a different situation to that of UK mainland areas. The effect of the ANM has primarily been to shut down renewable sources when the excess energy has threatened to short the connector cable with the national grid, which in its current form is unable to be smart.

Laura Watts's research on innovation in Orkney points to the specific intra actions of place, people, orientation, topography and technology research. She argues that:

> The futures of the renewable energy industry, here in Orkney, are an effect of this particular place. But to know rather than mimic them, you must dwell here. For knowledge is not exported but made and re-made where you are. Orkney is an island experiment in renewables future-making, a landscape that resists slippy quick-collaboration, you have to become a part of the experiment to make and know its futures. (Watts 2008: 9).

Watts makes the argument, following Haraway and Barad, that the future is situated. The future of high-tech imaginaries is 'Anyone Anywhere Anytime; Ubiquitous; Always On … in air conditioned rooms filled with telecoms switching equipment; a profit-making colonisation of people and place with the assumed moral authority of technological development' (Watts 2008: 2). However, the air-conditioned rooms with high-tech telecoms are situated in Silicon Valley or other company offices where this kind of future is made. This is not everywhere. The anyone, everywhere and anytime of the technological imaginary is not the same for anyone, everywhere and anytime.

In Orkney the topography, location and weather are rather resistant to seamless and ubiquitous connectivity, and the same conditions make the area rich in renewables. However, the ubiquity of climate change and its capacity to create new conditions and effects, together with crises in economy and politics, make these always-on mainland futures themselves seem less tenable. Even in Silicon Valley, the heartland of the ubiquitous silicon future, histories and futures of pollution, ecological damage, crises in resources, changes in weather, and seismic activity are very close to home (Anderson 2015).

In 'Electric Nemesis' Watts (2016) criticizes the lack of any reference to Orkney in the 2016 UK policy documents about smart grids. She notes the claims of revolutions and the demand that 'all change', noting that the call for total change obscures local successes that already exist. In the case of Orkney, the national network works against it because it is already configured as a peripheral node. In some ways Orkney would be better going off-grid, as the main reason its renewable sources get shut down is that the grid connection to the mainland can't handle the energy coming back:

> The grid cannot cope. The operator, Scottish and Southern Energy (SSE), have taken emergency measures and in 2012 slapped Orkney with a moratorium on renewable energy generation. No more wind, wave, or tide energy. The cable will melt and the entire electricity network will fuse, otherwise. (Watts 2016)

The flow of immaterial material, electricity, is formed as an industrial product of an energy industry infrastructure. Here, as Watts notes, the network infrastructure is a postwar grid disseminating electricity from power stations to energy consumers. Orkney is now an energy producer, ostensibly fulfilling the mission of the smart grid vision, but connected to an infrastructure which cannot cope. The all-change mantra of the Smart Power statements doesn't include a plan to replace the connecting cable with Orkney or to direct the excess renewable energy into a national grid structure. In the meantime, the energy industry in the UK is set on putting smart meters into millions of households, again and again.

When it comes to the energy futures of smart grids, antithetical devices materialize, together with hubristic documents and advertising images. In one version of reality, Orkney renewables and their Active Network Manager fall off the map. Margins and centres are always relative. Seen from London, Orkney is marginal, a group of islands off the northern point of a national geography which has prioritized the south and imagines London and the south-east as the centre. The smart grid imaginaries of the London-based government don't consider the energy futures of Orkney (Watts 2016), and step over them in an imagined relation to Iceland. In the Smart Power vision a possible connecting cable between Iceland and the UK is imagined instead. These centralized visions of smart grids fail to bring publics on board, partly because of their abstract rendering and lack of possible identification, and partly because of the way energy consumption has been denationalized and commodified. Like global warming itself, smart grids are challenging to represent in meaningful ways. Interesting, then, that a tangible set of experiences in Orkney goes unrecognized as a valuable resource, not just for the smart grid as a technological system, but for the smart grid as a possible experience. Although smart grids are unreal objects, they have material significance and practical implications for the actors engaged in crafting the vision, and in their development as technological objects. They promise to enact energy futures, while forgetting the past and present of an unevenly distributed, disconnected set of energy realities.

CONCLUSION: (RELATIONAL) MATERIALISM AND SMART GRIDS

Smart grids entail dirty objects as well as clean imaginaries, and they bring the industrial materials of telecoms together with those of energy. They bring together epic visions with small objects, dreams of using desert power, draining seas, putting carbon back somewhere in space and time. They enable seductive fantasies and real attachments, rapture and nostalgia. They provide an opportunity to consider why some speculative futures are taken up more readily than others.

Questions about how to value not just contrasting, but competing versions of reality are at the centre of this project about unreal objects. Power in the smart grid zone is at once productive of energy, politics and communications. In the case of smart grids a network imaginary is recapitulated as a grid imaginary with some kind of top-down control. The network politics of the information society are to extend to electricity. However, this is circular in the sense that electricity networks enabled computing infrastructures in the first place. Smart grids produce smart meters, government policy, innovation and shifting power relations. Actual experiments in renewables, off-grid projects, transition towns and other local energy collectives do not appear in the object of the smart grid generated by government or by the energy multinationals. However, these latter actors are the ones that get to say the most about smart grids. And what they say is largely materialized through media forms like adverts, policy documents and share prospectuses. These promote smart meters and other technological objects such as the prospect of building an infrastructure to link with Iceland, and meanwhile obscure the existence of other more local energy sources, like Orkney's production of over 100 per cent of its energy through renewables.

Under these conditions, stories about energy futures which computerize the energy grid, create new consumer electronics in the shape of smart meters, and promise to bury excess carbon underground seem like misplaced fictions. However, without significant intervention or changing conditions, these futures look more likely to materialize in the UK than those in which there is any genuine reduction in electricity consumption or significant shift to renewables. Electricity was made visible through the telegraph (Batchen 2006), and a world of electronic media has emerged as a product of an industrial model of electricity as a utility. The ur-media of electronic media, the computer, is undergoing a similar process of remediation, where it seeks to become a utility form, both medium and content. Networks and grids are powerful configurations. The question of whether the network mode will animate energy futures as networks of people, things and processes, as opposed to the grid as a network of objects, is an important one for everyone.

5

Real Fantasies:
De-extinction and In Vitro Meat

So far this book has examined the media processes and production around emerging science and technology projects which figure in the world as objects. The point has been to challenge a story of objects as they are given, to pull out some of the detail of how media realities are made and to look at some alternatives. The examples have been epic in terms of scale, investment and history, but those in this chapter contrast with these established forms. De-extinction and in vitro meat could be read as much smaller and more novel, the vanity projects of particular elites, and thus as less compelling in a story about technoscience, digital culture and society. However, like Elon Musk's visions for space travel, these projects have elite status and attachments in political terms. There are registers in which they are taken seriously, and their aspirations and proximity to ruling elites make them important objects to think with.

The examples in this chapter are approached to provide a different way into thinking about media materiality. They are framed here in relation to the theoretical language of biomediation (Thacker 2010; Kember and Zylinksa 2012; O'Riordan 2010) and rendering (Shukin 2009) in order to talk about biodigital objects. Eugene Thacker writes that 'biomedia are not quite things or actions but processes of mediation' (2010: 126). The previous objects – genomes, biosensors and smart grids – all generate new forms of representation that make a difference in the world. These forms of representation could be described as informational, where texts are taken as things in themselves rather than as forms of representation. The question of who gets to authorize meaning making is crucial in an informational culture where new forms of representation are not recognized as such. However, the examples in this chapter provide insights into what happens when those informational representations are taken as the basis for making new entities in the world. Thacker argues that 'biomedia present a view not merely of biological life as information, but of biological life that is life precisely because it is information' (2010:

126). The basis of information is a unit; to make something informational or subject to information technologies is an object-making process. Like turning things into data, making things informational can also be a kind of object making. The following examples are revealing about the kind of living objects made possible in object orientated registers.

De-extinction promises to make extinct animals alive again. It is offered in part as a biotechnological solution to climate change. It is about making new organisms and calling them old ones. In vitro meat is about taking the animal out of meat in what could be thought of as uncowing the burger. De-extinction and in vitro meat provide an insight into what happens when life that is digitized, or taken in terms of objects, is used as the model for new forms of materiality. The chapter explores how these examples are part of systems where digital and biological infrastructures (databases and bodies) come together in new kinds of circulation. It looks at the deconstruction of species, alongside the attempt to replicate and copy species, and attends to the space of flows as a metaphor for these new kinds of circulation. I use the chapter to build on the idea of informatic materializations, exploring species and materials to suggest that it would be useful to think about how the information politics of recursion allow new kinds of incursions in the real. The chapter looks at how fantasies are made real and the kind of work that has to go into maintaining and remaking these as objects. It brings questions of materiality to the fore by looking at data–flesh relations more closely and considers the interplay between rendering bodies as data in the case of genomes, on the one hand, and rendering bodies from such data on the other.

Another register in which to express this is the language of wetware, which suggests an articulation of how texts, code and software flows into biological materials and wetware. In the idiom of wetware, rendering is a useful term to think about how things are built up and simultaneously deconstructed. The idea of rendering is drawn here in part from Nicole Shukin's (2009) suggestive use of this term to examine animals and film. Rendering is also used in the register of digital image production, and graphics. It bridges the biological and the textual. It both builds up, in the sense of making things and making up graphics, and breaks things down, as in rendering fat from animals (Shukin 2009: 49). In these examples, bodies are media forms: they are rendered through a complex interplay of mediation and materialization, expressing conditions of reductionism, technological utopianism and economic-centrism as well as manifesting anxieties about the past and the future. They are symptoms of a world of TED thinking (both have featured as TED content), of the love of technology, of the seduction of the tale that technology will save the future and thus provide reparation for the past, and paradigms in which the conditions of possibility are informatic, molecular, digital and dividuated ways of understanding life. They are about species, doing and undoing, and the rendering of fantasy as real.

RENDERING IN VITRO MEAT: UNCOWING THE BURGER

In vitro meat is the making of meat in laboratory conditions, outside of the animal. 'In vitro' means in glass, but through the use of the term in vitro fertilization it has become synonymous with the laboratory or test tube. In vitro meat derives from practices of tissue culturing and engineering in which tissues are grown for the purposes of creating animal tissue for consumption. It is different from protein-based foods or mock-meat substances. These foods do not contain animal protein, are meat-free, and are marketed as such. In vitro meat is specifically about growing meat outside of the animal and it is marketed as meat. The product is created through cell culturing, or the growing of cells and tissues for medical research. These are relatively common practices in biomedicine. Stem cells, organ growing and skin grafts are all areas in which tissue culturing occurs. In vitro meat is at a tangent to these practices and offers a new pathway for tissue engineering.

In vitro meat has a high media presence, which peaked in 2013. It has materialized in the twenty-first century through a series of events, including engagements with art, public experiments and press releases. Throughout the late twentieth and early twenty-first century it was an 'as-yet undefined ontological object' (Stephens 2010); today, it still retains what Neil Stephens (2013) refers to as 'ontological ambiguity', although a major promotional event in 2013 did much to stabilize it as edible meat, and specifically as beef. This occurred partly through the staging of the cooking and eating of in vitro meat as a cultured beef burger in front of a live studio audience, and through the inclusion of one of the burgers created for this event in the Boerhaave, the Dutch Museum for the History of Science and Medicine. The proof that it was a burger, edible and with a meaty texture, was in the eating (O'Riordan et al. 2016). Its capacity to be ingested by the body and digested and broken down again, its destruction, was proof that it was a real thing, while its transformation into an inedible archive as a permanent museum object in the Boerhaave also testified to its reality and stability. Since then it has been taken up as the basis of a Silicon Valley start-up – Memphis Meats – this time in the form of a meatball.

Protagonists of in vitro meat have framed the idea of growing meat outside of the animal as a technology that might be able to alleviate animal suffering, feed large populations and reduce global warming. So far in vitro meat experiments have been conducted by the US space development agency NASA using fish cells; by the bioart organization Symbiotica using frog cells; and by the in vitro meat consortium, specifically Professor Mark Post, using cells from cows. It was Post who created the cultured beef burger cooked and eaten by food critics in front of a live audience in 2013. He referred to the burger as 'meat but not in a cow'. Overcoming the yuk factor and steering a public response towards affirmation and approval, the event marked a moment in the materialization of a new kind of flesh which also offered a break with previous incarnations. The cultured beef burger was a distinctive object in conjuring a sweeping vision of change and innovation in which saving the world from environmental challenges was invoked. However, like other innovations, in vitro meat is producer driven, an object looking for a cause, an innovation searching for take up, a product looking for a market. Like other emerging technologies, the promotional materials and key actors in the field link the technology to climate change. In this case, world food distribution and the challenges of feeding an overpopulated world are also evoked.

Making in vitro meat requires originating cells, so some cells from an animal are needed to start off the culture, and to feed it. However, once the tissue is cultured a potentially indefinite amount of cell production could be generated from this process. Hence Post's claim that cultured beef is meat without cows. Since cows here stand in for slaughter, the message is that cultured beef is meat produced without killing cows but still with living cows. This is a bit disingenuous not least because tissue cultures require foetal bovine serum (which has to be repeatedly made from dead cows) to grow the cells. However, as with other entities on a tissue culture spectrum (Mitchell and Waldby 2006), the boundaries of life and death are thrown into question by in vitro meat.

Symbiotica, the Australia-based art science group, have used in vitro meat in several of their projects. In an exhibit in 2003 they used cells to create frog legs, and then ate them in an exhibition space surrounded by the live frogs from which the cells had been taken. The promise that you can have your meat without slaughter has been taken up by the organization People for the Ethical Treatment of Animals (PETA). PETA have endorsed the cultured beef burger, offered a prize for in vitro chicken production, and sponsored postgraduate research into the area. The in vitro meat consortium that has emerged in this area have framed in vitro meat as 'an innovative technology offering solutions to world problems, identified as food production, population growth and climate change' (O'Riordan et al. 2016).

The framings of in vitro meat have shifted over a short time-scale from 2000 to 2013. Stephens notes that one of the most marked shifts in its promissory narratives is the move from the promise of providing food for space travel to that of addressing climate change. At the start of his fieldwork the space-food narrative dominated talk of the purpose of in vitro meat. However:

Ten years later, only the original group working on the NASA project included space travel in their rationalisation for in vitro meat. Newer entrants to the field, including Post, New Harvest and PETA, reconfigured the imaginary around the environment, animals, health, innovation and profit. (O'Riordan et al. 2016: 5)

The 2013 launch of the cultured beef burger was a hybrid event, the entirety of which was packaged as a live stream and picked up by television news as well as through direct viewing. The studio event in which the cultured beef was cooked and eaten was followed by a question and answer session with the audience. The event was presaged by the screening of a promotional video, produced by the creative agency the Department of Expansion. The video featured talking-head shots with Sergey Brin, Richard Wrangham, Ken Cook and Mark Post. Brin was the investor, Wrangham provided a commentary from an evolutionary biology point of view, and Cook represented food science. The talking-head shots were interspersed with images including video footage of the Californian coast, images of ranchers in the US and intensive farming, visualizations of ancient humans eating meat, followed up with men barbecuing meat in a modern suburban back yard.

The video as a whole promoted in vitro meat and these key actors as part of a world-saving project in sympathy with human evolution. Intensive meat farming was represented as a major problem that now threatens humans because of environmental damage, as well as an issue of cruelty to animals. Meat consumption was represented as a natural and necessary feature of human evolution, which now threatens humans because of intensive production methods together with rising demands for meat. Ranching, hunting and cultured beef were situated as natural modes of engaging in meat consumption, and cultured beef in particular was offered as the rational solution to the problem of providing enough meat without destroying the environment.

The screening of the promotional film (available at culturedbeef. net) was followed by the live studio event. This was located at a London studio and a selected audience was invited to witness a celebrity chef, Richard McGeown, cooking the burger and the subsequent tasting of the same by two food critics, Hanni Rutzler and Josh Schonwald. The event was hosted by news anchor Nina Hossain, who narrated the proceedings, interviewed Mark Post about the production of the burger, and the chef and food critics about the cooking and taste. Rutzler and Schonwald emphasized what they called the meatiness of the mouth feel, and confirmed that it was edible and similar to a burger. The studio audience followed up with questions, and made something of a protest that none of them were able to taste the burger.

Science media in the UK and globally reported on the event, and key protagonists weighed in with supportive interviews and editorials. Isha Datar, from the in vitro meat lobby New Harvest, was in the audience, and curated a Reddit thread about the event afterwards. The campaign generated a high media presence for the event in the days immediately before and after. There were over a 1,000 posts on the Reddit thread while Datar was online, and clips from the event were picked up on global television news networks. There were also reports in the *New York Times*, UK broadsheets and key European newspapers.

The event was part of a tradition of creating scientific knowledge via press releases (Haran 2007). This contrasts with a scientific history of peer review and publication via journals, but is consistent with a history of public experiments and the witnessing of demonstrations as a mode of knowledge making. This mode is also consistent with the managing of media forms as sites of public engagement with science (Haran and Kitzinger 2010).

Publics for science and technology have been cast as excited and awed spectators, such as the live audiences for rockets taking off at Cape Canaveral, and the television audiences for the moon landing in 1969.

They have also been cast as witnesses in validating public experiments. For example, when the Large Hadron Collider was switched on, and in subsequent events, CERN invited groups of journalists and students to witness the events and report back on them to their communities. However, the use of creative agencies and the staging of media events around press releases is slightly different to this. A history of public engagement via press release has been attached to maverick science such as reproductive cloning (Haran 2007). It has attracted negative attention as well as media exposure. For example, the use of high-end video material to provide rhetorical force and spectacle, and the staging of an experiment as a reality television show, cast the cultured beef event in an uncertain light. The launch negotiated this space: on the one hand, it allowed the burger to be undermined as science because it didn't go through peer review; on the other hand it allowed it to be framed in terms of spectacle and media authority through a well-considered campaign. What is different in this moment, and makes this space more easily appropriated than it could have been only ten years ago, is that social media has become the form through which news is consumed and circulated. In this mode it is not peer review or gatekeeper authority that determines knowledge value, but attention, number of clicks and circulation that determines whether something counts as core knowledge. In these terms the event was successful in generating media coverage and interest and in deflecting substantive criticism or antagonistic publics.

Since the 2013 event, the California-based start-up Memphis Meats, backed by New Harvest, has begun to explore culturing pig cells to make meatballs. They have also produced a promotional video of the cooking and tasting of a cultured meatball. Some of the images in their film referred directly to the burger press release, which featured close ups of the meat cooking in the pan and commentary about smell and taste. However, the Memphis Meats film is hard to take seriously because of its potential as satire, and its nods to the cultural beef material are much more humorous. It lends itself towards a parody of the genre of start-up promotional videos. At the end a young women is filmed eating the product, telling the audience that it tastes just like a meatball. The production values are much lower than in the cultured beef launch, and the profile of Memphis Meats is much less distinct. However, it points to the extension of interest around the product.

An extended Reddit thread curated by Datar in early 2016 linked the Memphis Meats start-up to the legitimacy of the earlier event. In the discussions the main obstacle to in vitro meat was framed as psychological, in terms of the yuk factor, or irrational fears of chemicals and unnatural products. There was also some consideration of how to move in vitro meat further away from the body of animal. Part of this was based on the assumption that foetal bovine serum is currently required as part of the culturing process. Datar suggested that the genetic editing technique CRISPR/Cas9, which is also implicated in cloning, might offer a solution to this. Although she was vague about how this might be realized, it is striking that CRISPR gets pulled into such a wide range of areas at the moment as offering a possible solution. It has become a new object to think with, linking multiple projects and providing a connecting node in joining up media commentary. Key terms, especially ones from the biosciences literature, also operate as nodes in this area creating discursive fields which help to stabilize this kind of phenomena. The structure of social media filtering algorithms means that once someone is looking at in vitro meat, cultured beef or linked sites, more such sites come up in searches and news feeds, reinforcing its legitimacy and making it seem more real.

During the burger launch in 2013 the participants claimed that the project to produce enough tissue to create two burgers cost over £200,000 and took over two years to complete (although the final tissue was cultured over three months). This price tag cuts out the years of research and development and externalizes the earlier work that led to the point at which the project could be started. Thus, the promise of in vitro meat is emphasized while the costs of producing it are diminished.

In vitro meat is a hybrid network bringing together people in the art world, tissue engineering, public relations, Google, food criticism, chefs, sociologists and animal rights organizations. The networks of nonhuman materials range across nonhuman animals, art works, engineered tissue, public relations materials, journal articles and the apparatus of PR events, laboratory experiments, and a science museum.

In vitro meat can be seen as an extension of both animal capital and the rendering of the animal (Shukin 2009). Shukin argues that animal capital is a literalization of the commodity form, where animal life itself becomes economic life, a form of money. Animal life in the context of in vitro meat is, as Catts notes, disembodied, although not entirely. Animal life here becomes generative of animal capital, which appears as if it could exist without animal. The value of the animal as food is attached to the object created to make this meat.

When human cells are cultured in tissue engineering, and particularly in the generation of stem-cell lines, foetal bovine serum is often used because it promotes embryonic growth. Thus, human cells created for biomedical intervention are commonly mixed up with nonhuman animal cells (M'Charek 2005), and hybrid human-animal embryos are offered for research (Haran 2011a). However, in the case of in vitro meat, the purity of the animality is insisted upon through the representational choices made in the promotional materials. For example, Richard Wrangham claimed in the cultured beef promotional film that humans were evolutionarily dependent on meat eating, always represented as nonhuman animal meat through visualizations of an historical imaginary of hunting animals, and a contemporary imaginary of the hunter-gatherer tamed in the image of men cooking on back-yard barbecues.

The cultured burger was thus presented as a straightforward progression in meat history, offering a more modern form of meat. This emphasis on real meat offered a distance from mass-produced or fast-food meat, with its connotations of being mixed with horse meat or rodents. In the online comments around the burger launch various people tried to insert comments about cannibalism or pose versions of the following question: if there is no cruelty or violation of rights involved, why not culture human meat for consumption? However, the authoritative actors in these threads always led the discussion away from these topics, either by ignoring them and letting them sink into invisibility, or by steering the discussion firmly back to cultured beef as a form of nonhuman animal meat. Beef and chicken are the currently favoured candidates for in vitro meat production, with the burger as the main contender for the form it could take as an edible mass product.

In vitro meat expresses a kind of digital way of knowing, or biomedia, as well as discursively traveling through artefacts of digital media. Thacker notes that 'there has been relatively little exploration of the ways in which an informatic paradigm pervades the biological notion of the body and biological materiality itself' (2003: 48). In the years since he published that piece the pervasive influence of the informatic paradigm has been explored in a number of ways, but its authority or homogeneity has also been challenged. Mitchell and Waldby's (2006) work on tissue economies and Sunder Rajan's work on biocapital (2006) follow the complexity of these flows at the level of political economy, tracing who profits and who doesn't in the global markets of biomedicine.

Art practice has also offered a substantive critique of forms of biomedia (De Costa 2008). In Symbiotica's earlier work with frogs, for example, they framed tissue culturing and in vitro meat as 'disembodied cuisine' and 'semi living' (Catts, undated). Catts describes these terms as 'less scientific but more fitting', which speaks to the contestation over meaning in this area. Catts, the Tissue Culture and Art Project, and Symbiotica have produced as much in vitro meat as the in vitro meat consortium, and they are significant participants in making this object. This points to the way in which, in some cases, art that links to science is not about artists responding to or illustrating science but about them becoming implicated in making the scientific field. Bioart, which takes the practices of the biological sciences and the materials of biology as its medium, is implicated in tissue culturing, and is part of the genealogy of in vitro meat. At the same time, the work of Symbiotica brings the mixing up of media production and biological production in the process of making in vitro meat much more clearly to the surface. However, even after more than a decade of heightened attention to biopolitics and information politics since Thacker's observation, it is still often the case that discussions of biopolitics background informatic infrastructures, and discussions of information politics background biotechnological structures. Biological samples pass through databases, and these repositories are forms of life (Thacker 2004; Zylinska 2009) and of corporealization (Haraway 1997; Mackenzie 2003). In the end, however, most media forms are forms of life and should be taken as such.

RENDERING DE-EXTINCTION: SIMULATION AND THE CULTURE OF THE COPY

De-extinction is the project to bring back extinct animals through a number of technologies including cloning and genetic editing, as well as cross-breeding with contemporary species. It is also linked to climate change amelioration through the idea that it might restore habitats as well as offer reparation. There are clusters of scientists world-wide invested in this project, and it too reached a peak in popularity in 2013. This year saw a series of experiments, books, TED talks and other forms of dissemination bringing the subject to public attention. De-extinction requires viable ancient DNA from the extinct species targeted, and a close contemporary species to host a cloned embryo or fertilized egg. While the mammoth is probably the most charismatic candidate species in these networks, the extinct North American passenger pigeon is also a favourite. Viable DNA from the passenger pigeon exists in museum and biology collections. Other pigeon species exist which potentially enable the fertilization of live eggs.

De-extinction has a high media presence but for the moment at least it is an ontological impossibility. A cloned hybrid might materialize an approximate phenotype of an extinct animal, in so far as that can be judged, but this is not the same as the thing itself. Like other forms in this book, the question of the object is deferred as it is made. Materializations of de-extinction to date include its presence as a textual trope, the materialization of media materials, genetic engineering experiments, and cloning experiments which materialize organisms. For example, the live birth of a cloned Spanish Ibex in 2009 was created with DNA from the extinct animal and the ova of domestic sheep (the animal died shortly afterwards). The claim that the Ibex was an example of de-extinction privileges nuclear DNA as the carrier of the property of species and underplays the significance of the egg and mitochondrial DNA. In order to claim that this was a case of de-extinction the materiality of the egg and its DNA have to be excluded from any definition of the identity of the offspring, so the extinct nuclear DNA can be used to claim the identity of the live birth.

De-extinction, then, both defines and destabilizes species as a category by excluding some materials like those of the ova or egg, and by writing identity into the nucleic DNA only. It does this in a mode of disavowal, or the fetishization of the nucleic DNA as species. Such examples of de-extinction could also be thought of as cloned, hybrid organisms or genetically engineered new species, but they are framed in terms of the identity of the extinct species. Thus, the language of de-extinction is a form of purification, making up the identity of the offspring of biotechnological intervention as an origin story or a return to a previously natural state.

The *Jurassic Park* film franchise has given de-extinction a high media presence, and the explosion of merchandising around the film (Franklin 2000), as well as the popularity of special effects in documentary forms (e.g. *Walking with Dinosaurs*), has given extinct animals a symbolic currency in terms of forms of realism. The perception that, through these representations, people in the present have seen what dinosaurs actually looked like in the past is pervasive. Once the media of representation is that of the animal, rather than that of the filmic text, then representation and actualization are collapsed. Thus, both textual representations of de-extinction (graphic rendering) and the biotechnological rendering of the animal are simulations of species and come together as de-extinction technologies. Media production and biotech come together in conjuring objects. These entities have to look, feel and taste like existing things to count as real, so the aesthetics of media making and those of biotechnological rendering are caught up together.

De-extinction promises more fleshly realities from digital cultures and informatic paradigms. It makes animals instead of undoing them. De-extinction doesn't have a single media event like that for the in vitro meat burger, but a series of events and objects come together to make it present. These include Hollywood films, press releases, TED talks, books and book tours, a celebrity science culture, museum exhibitions, social media, newspaper articles, TV features, cloned organisms, and experiments in genetic engineering. To date there are very few claims to have materialized de-extinction. However, there are a vast array of media materials illustrating, exploring and promoting it.

In 2013 *National Geographic* ran de-extinction as a cover story and also hosted a TEDx conference on the topic. The Revive and Restore foundation was at the centre of this, and brought together key figures in the area. Like in vitro meat, de-extinction brings together a fairly small group of people but gets high-value coverage. The Harvard geneticist George Church is a key figure, as are Ryan Phelan of Revive and Restore (also previously of DNA Direct) and Beth Shapiro, an evolutionary biologist at the University of California Santa Cruz. Another key actor is the Pleistocene Park project to restore the Mammoth Steppe Ecosystem in northern Siberia. While the project hasn't materialized mammoths, it is a conservation project working with existing species to re-create productive pastures with high animal density.

The most recent iteration of the Jurassic Park franchise, the 2015 release of *Jurassic World*, reaffirms de-extinction's seductive promise and threat through the mega-spectacle of dinosaurs on screen. The first film in 1993 prefigured current ambitions to clone the woolly mammoth. The fauna of de-extinction range from charismatic mega fauna to modest proposals. Among the other candidates promoted by Revive and Restore, the much more modest passenger pigeon is the frontrunner. Driven into extinction at the start of the twentieth century, in narratives of its demise it symbolizes the march of progress coupled to the destruction of nature, standing in for industrialization and colonization. The de-extinction of the passenger pigeon promises a return to the lost environments and species of an earlier United States. It represents a modest attachment to the unreal promise of de-extinction. The fleshly realities promised through de-extinction offer new species through cloning projects in another guise.

In common with in vitro meat, while there are published scientific peer-reviewed papers, many of its big moments have been public announcements and media events. There is a media culture of de-extinction through which ideas and research circulate. In 2015 Church claimed to have engineered functioning elephant cells with mammoth DNA. He gave interviews with journalists and made announcements at scientific gatherings but did not publish the research. Shapiro's tour for her popular science book *How to Clone a Mammoth* (2015) also received media coverage and led to speculative and entertaining representations on the subject. Some coverage in this area refers to the Jurassic Park franchise as having a negative effect (Zimmer 2013) because the film invokes a vision of technology out of control and on the rampage, figured through the dinosaurs. The film is cast as an obstacle to broader acceptance and support.

In fact, the first Jurassic Park film was tied up with the promotion and development of the field of ancient DNA. The consultant on the film – Jack Horner – was awarded a grant for ancient DNA research, and later research on de-extinction was funded by George Lucas (Kirby 2011). Other findings in ancient DNA research, published in *Nature*, were tied into the release dates of the film because of the promotional value of Hollywood. Hollywood has often acted as a vehicle for promoting areas of science or securing investment in prototypes (Kirby 2011), even when its film plots draw on tropes from horror or dystopian science fiction (O'Riordan 2010). Rather than casting a shadow over de-extinction, the Jurassic Park franchise has generated investment in the field, facilitated a broad social understanding of de-extinction, and added to it the seductive spectacle of the dinosaurs.

The most dominant way of making de-extinction legible is through the use of illustrations of extinct species to conjure an imaginary of return as well as to memorialize extinct animals. The mimetic realism of these images is powerful and often taken as given. There is a history of contestation as to what dinosaurs really looked like, for example in debates about whether they walked on two or four limbs, or whether they had feathers and wings or scales. However, specific images of dinosaurs rendered in 3D through computer animation have become taken as accurate representations. Likewise images of the mammoth or the sabre-tooth tiger have become iconic to the point of clear legibility and comprehension in visual terms. Despite never having existed in the lifetimes of anyone alive, or, in the case of dinosaurs, never in human history, we have a series of representations in which these species have a visual life. Extinct animals, of a particular charisma, have a disproportionate visibility given that they have never been seen alive. There are of course visual sources of evidence for such representations, the fossil record being the most significant, but tissues preserved in ice and amber have also contributed to visions of dinosaurs and mammoths.

Species that have become extinct during recorded human history have a different provenance in that they have often been preserved through taxidermy or other museum archiving processes. They were also the subject of illustration and later photography, and are inscribed in texts which still circulate. The Revive and Restore website reproduces a painting called *Gone*, a composite image of some species that have become extinct

Figure 5.1 Homepage image on the Revive and Restore website (reviverestore. org)

since 1700, on its homepage (Figure 5.1). This 2004 painting by Isabella Kirkland was produced from museum samples and natural history illustrations and is a striking testament to the last 300 years of practices in natural history as well as to the 63 extinct species portrayed. The painting shows fragments of bone, feathers, preservation jars and taxidermized animals and reproduces other painting and illustrations. It reminds us that vision is a form of knowledge making and one that dominates in the history of science (Cartwright 1995). What is produced as visual evidence in scientific knowledge offers a form of reality. The historical provenance of vision in this case provides an assurance that these really are the extinct animals. However, the compelling power of spectacle in the case of the dinosaurs and mammoths, and the high technics of CGI, provide another kind of assurance. The work of visualizing extinct species is a form of reparation, retrieving what has gone. The evocation of absence and the emphasis on anthropogenic extinction in these projects makes de-extinction a reparation story, even though its most charismatic species pre-date human intervention.

Cloning is an important node in these new imaginaries of reparation. The conditions of mammalian cloning were made possible by the flow of eggs outside of bodies. Thus, technologies of breeding in mammals underpin the development of in vitro fertilization (IVF) in humans, where disembodied or in vitro techniques for fertilization were first developed (Franklin 1997, 2013). IVF was developed as a technology for human fertilization in the 1980s, and one product of this field was the circulation of human eggs outside of the body for the first time, making them available for laboratory use. Thus, IVF opened up human cloning in a process that Karen Throsby (2004) refers to as 'technology creep'. In the UK, new legislation in the 1990s and the early twenty-first century established a legal framework for human cloning, hybrid embryos and genetic selection and modification. Just as IVF created new realities such as cloning, so cloning has opened up the path to de-extinction in another instance of technology creep, together with a wholesale reframing of legitimacy. Cloning has been rebranded as de-extinction by some actors in the field, for example Beth Shapiro claims in an interview that scientists 'brought Dolly back' because the sheep from which the donor cell came was dead by the time of the cloning experiments. At the time Dolly the sheep was cloned there was no framing of this as de-extinction, and indeed any allusion to tropes from horror, science fiction or religious themes such as resurrection were firmly eschewed in the framing of cloning as a modern and hopeful biotechnological success story (Franklin 2007; Haran et al. 2007).

Some 20 years later, Dolly is making another come back, but now as the poster sheep for de-extinction. Claims such as those by Shapiro build on the success of the Dolly story as hopeful modern science, but bring back tropes of resurrection and the undead. Twenty years on it seems that even zombies are available for co-option in the optimistic discourses of biotechnology. De-extinction and cloning are derived from the same biological techniques of somatic cell nuclear transfer and genetic modification. However, they also draw on assumptions of similitude, or likeness and sameness. Cloning and the digital culture of the copy and simulation are very much mixed up (Stacey 2010; Munster 2011; Schwartz 1996; Battaglia 2001). Copying, cloning, three-dimensional rendering and animated simulation have been possible in digital image composition for much longer than they have been in biotechnology. However, the conditions of the one also make the other possible, for example when the simulation of dinosaurs on screen prefigures the possibility of cloning them off screen. Both forms of copying require extensive labour to render similitude (Munster 2011).

De-extinction imaginaries reconstruct species as the object to be copied, and deconstruct species stability as a possibility through genetic engineering. The slippage between the two facilitates the idea that de-extinction is possible, but it is a contradictory narrative as many of its protagonists admit. Shapiro, for example, acknowledges that cloning mammoths and de-extinction are not possible. She negotiates a complex rhetorical space, claiming in her press release that de-extinction is possible, while arguing in greater detail in the book itself that it is a form of cloning that could never be a true species return, but will always be an approximation of a best guess, plus whatever DNA is available. In any possible de-extinction future the egg or ova of a host species for the clone would be necessary for a live birth, or hatching, so there would always be a mixing up of DNA. On the one hand Shapiro attributes belief in de-extinction to a sensation-loving media culture, while on the other hand she promotes it as a brand through extensive discussion in her book.

Shapiro argues that de-extinction in relation to ancient DNA could never be actualized because the genetic materials are not available. To tell a story about why species are not species and how this all goes back to IVF and cloning it is necessary to take some steps back. Coming close to creating something that looks like a mammoth would require inserting nucleic DNA from a mammoth cell into the de-nucleated egg of its nearest genetic relation, the Asian elephant (itself an endangered species). A resulting embryo would have the nucleic DNA of the mammoth, with the mitochondrial DNA of the Asian elephant. George Church's team have already decided what the species-defining elements of a woolly mammoth are. The definition includes blood oxygen release at low temperatures, fat storage and hair. There is insufficient ancient DNA to import mammoth DNA into an egg, so his team are using CRISPR genetic editing techniques to engineer Asian elephant cells to create cells that can code for haemoglobin, fat and hair to synthesize mammoth DNA for the cloning described above.

Another register for talking about de-extinction, then, is that of cloning and genetic engineering. Species definition is an important element in the space of materials flowing around de-extinction because it anchors the language of de-extinction, steering it away from cloning and genetic engineering, or the even less popular registers of genetic mutation and genetic modification. A long history of media representation and public engagement with genetically modified crops has stabilized the meaning of GMO as largely negative, at least in Europe, and it is noticeably absent from the register of de-extinction. Species becomes an object to think with, simulate, animate and materialize through the biomedia of de-extinction.

FLESHY NODES IN THE SPACE OF FLOWS

Dyer-Witheford argues that the infrastructure of cybernetic capitalism constantly opens up new ways for capital to flow, in a cycle of infinite production and appropriation. However, along with other post-, neo- or autonomous Marxist approaches to media, he also maintains that there are spaces within the machine or total system in which new identities or resistance can emerge (1999: 227). Eugene Thacker argues that biomedia enables the same cycle, but also opens up new flows where the meanings of bodies, capital and machines change as information changes substrate, from wetware or bodies through databases and sequences, through to new iterations of wetware. The capital flow of biomedia thus also enables spaces of possibility and change. Thacker's sense of biomedia enabling changing meanings opens up an escape route from Dyer-Witheford's total machine. This sense of biomedia making a difference or changing meaning as patterns of information change substrate can also be taken in conversation with Tim Jordan's argument that recursion is a defining force of information politics. Jordan's take on recursion is drawn through its use in computing, and like Kelty's (2008) use of the term, also through its use across sites of popular culture. Jordan defines recursion as the process of making a difference to the same. Making differences to the same allows an indefinite control over the production of value because the same materials can be endlessly recycled – information doesn't disappear when it is passed on, it is endlessly copied.

In vitro meat and de-extinction destabilize their emergent objects, even as they materialize, and in this process make differences that move in repeated patterns of emergent stabilization, destabilization and disintegration. In vitro meat keeps on culturing cells so that they grow, only to be eaten. This is a process of forms of information not just eating themselves but becoming food. In vitro meat opens up a new site of circulation for animal capital, and de-extinction points to the forms of circulation through tissue economies in a digital register. De-extinction currently has two modes in the scientific mix, one is through cloning and the other is through genetic editing technologies (such as CRISPR); these can also be combined, since genetic editing doesn't preclude cloning. Neither of these modes would bring back a species in the way that our cultural narratives about species and extinction support. However, this too has recursive force: both cloning and genetic editing refer to themselves in their processes and de-extinction is analogous to parallel mirrors or visual recursion, in that the image of the species to come is the image of the species gone.

In media studies it has long been the case that media is understood as constitutive of life. This can be traced in different ways through the literature. For example, a ritual understanding of media use as everyday life foregrounds the role of the media in making the realities of everyday experience. A representational paradigm where identity is formed and negotiated through media texts and discourses also takes identities as made in the media. A socialization paradigm takes the media as the primary agent in shaping what we know of and how we understand the world. Through McLuhan, Steigler and media archaeology, media technologies are the constitutive prosthetic of the possibility of the human. Being human is always a communicative, tool-incorporated ontology in these fields. However, after biopolitics there is an understanding of a bios of new media where media texts are incorporated as bodily tissues, sites of flow between body and flesh. This is perhaps encapsulated by Joanna

Zylinksa's formation of blogging as bodily expression, drawn from Foucault's argument that writing makes flesh and blood, reformulated by Zylinska (2010) as: 'if it reads it bleeds'. Media forms create and capture affect, manage labour, produce identities, capitalize life. At the intersection of biopolitics and the informatic a prosthetic relationally opens up human and nonhuman animals, life and materials, to a site of exchange, and exchange value, capitalizing on and reshaping both the biopolitical and the informatic.

In vitro meat, in rendering animal capital outside of the animal, is an example of a medium flow in the same way that digitizing the genome allows the movement of capital across biological and informatic boundaries. The removal of meat from the body and its relocation in vitro enables a new flow, or circulation, of animal capital. Rather than the insane conditions of industrial farming becoming the point of friction against which the economic rationalization of life itself might stumble, in vitro meat provides a new opening for the recapitulation of animal life as capital. The semi-living matter of in vitro meat becomes a new medium for circulation. One of the concerns in environmental assessments of in vitro meat is that the culturing of meat, outside a body, removes it from the circulation of other living forms, of viruses, microbes and antibodies. The pristine conditions of in vitro might also open up new flows that are beyond the system of control; in Serres' (1982) terms: without the parasitic there is no life. This possibility of rupture is a similar opening to that made in Bardini's *Junkware* (2011), and to a certain extent in Shukin's *Animal Capital* (2009). That is, that the master narratives of control create their own contradictions, as the sterilization of in vitro meat production might produce unforeseen parasitic or biotic life, and its own food scares. The symbolic production of in vitro meat, as just meat, also produces its own contradictions, and the discourses of cannibalism and disgust cannot be cleaned out of the responses entirely. The current rebranding of in vitro meat promises that we can save the future and have our meat and eat it, but its instability also portends a perilous proximity to dystopian worlds of overpopulation, climate disaster and out of control capitalism.

MEDIA AND INFORMATIC MATERIALIZATIONS

One way to think about these examples of unreal objects is in terms of media presence on the one hand and ontological stability on the other.

If ontology refers to a category of being in the world, it is media life that helps to stabilize this reality. For example, in vitro meat has a fairly high media presence but lacks ontological definition, or in other words: WTF is it? Much of the making of in vitro meat consists of attempts to create ontological definition, and as we have seen it has oscillated in form through frog legs, fish and beef. Currently it has temporarily stabilized as an edible beef-derived product. It is the making of the burger into something that looks and tastes like a burger that provides this intelligibility in the world. That making took a lot of media work and the production of the object occurred through a media event.

De-extinction on the other hand hasn't been made in the world as such, in that there are no de-extinct animals alive. It has some media presence, through publicity material, popular science books, TED talks, coverage of media events, and a history of Hollywood spectacle, but it is at the same time an ontological impossibility. That is to say that, even were a de-extinction project to be played out, its ontology would be that of cloning or breed husbandry, not of de-extinction – nothing is actually coming back from the dead. However, through the media production of dinosaurs in film and documentary there is something like a trace of de-extinction already in the world that helps fuel the claims from the scientific community that it is an imminent reality. For, example when the iMAX Cinema at London's Science Museum first screened the 3D documentary *Dinosaurs Alive!* in 2007 this showcased the projection technology so that the dinosaurs appeared to have presence in the auditorium. For those of us there it did feel as though something of the reality of dinosaurs had been experienced, and for a couple of years afterwards my daughter (who was six at the time) still maintained that she had seen a living dinosaur.

Many of our emerging technologies, in the current moment, represent and constitute life in an informatic mode. From human genome sequences to a computerized planet through the Internet of Things, bodies are caught up in these informatic modes. In this modality the processes of mediation with which lives are made drop from view as representational and instead are claimed as informational, fact-giving, material making. A range of theoretical frameworks have been developed to address the politics of these conditions. Haraway (1992) developed the language of diffraction to talk about this, and develops the idea that databases are modes of corporealization (Haraway 1997). Adrian Mackenzie (2002) references these ideas in his work on transduction and corporealization.

Jordan's (2015) thesis on recursion also takes a line through the question of how a mode of information structures the conditions of politics. Both Mackenzie and Haraway point to databases as obligatory passage points through which information is corporealized. De-extinction and in vitro meat provide fleshly realizations of information where bodies provide data for informatic forms, and the substance of new fleshly entities embody the mode of information. However, they also provide sites of friction and resistance because these moves are not seamless and require great discursive effort and huge production efforts to appear so. Both the appearance of dinosaurs in the iMAX and the creation of a cultured beef burger demanded a huge, complex and expensive apparatus of production. Unpacking some of those efforts allows some of those frictions and contradictions to come to the fore.

The biomedia of de-extinction creates a space of flow for the transmission of the expression of species itself. Species as a concept, which goes hand in hand with ideas about biodiversity, is activated in the de-extinction discourse as something that can be animated and, through its animation, defined. To draw on Shukin's language of rendering, species is rendered through de-extinction and the figurative is materialized, or, as she says of the commodity fetish rendered as animal, literalized. The register of materialization and animation works effectively in relation to de-extinction – which if it were literalized would be a resurrection technology. However, while de-extinction could materialize, create or render a mimesis of species, it can't literally happen. De-extinction is a representational technology in that the goal is to create animals that look like their extinct counterparts, and have phenotypical properties such that they appear to embody the extinct species, even if the genetics are not the same. In this way there is an assumption that phenotype might rewrite genotype. Some of the cultural critique of DNA and its 'poetics' (Roof 2007), imaginary (van Dijck 1998) or code structure (Bardini 2011) focuses on the way that the Central Dogma and genetic determinism have been interpreted in popular culture.[5] De-extinction, like many genetic stories, both reinscribes genetic determinism and challenges it at the same time. On the one hand the idea of de-extinction hinges on the idea that nucleic DNA is powerful enough to determine the

---

5 The Central Dogma was coined by biologist Francis Crick (1958) and claims that information passes from DNA to proteins but not back. This has been interpreted in general terms as genetic determinism, or to mean that DNA determines biology.

expression of the organism, on the other hand it allows that phenotypical appearance might determine identity.

Images of extinct species are important in the circulation of de-extinction materials. The project promises to bring to life things only seen in representational forms, across media texts from historical natural history illustrations and photographs or documentary footage, to simulations of animals never filmed. Established genres of documentary simulation come together with the imaginary of biological simulation such that success in the latter would demand a correspondence in the former. In other words, the success of de-extinction would require that the candidate species looked like the documentary evidence for the species likeness or similitude. However, species recognition and similitude have to match.

Another way of thinking about de-extinction and species is through much earlier projects to bring species back. In Jim Endersby's book *A Guinea Pig's History of Biology* (2007), he tells a story about the quagga, an extinct sub-species of plains zebra extinct in the wild by the late 1870s. Endersby relates the story of an attempt to restock the quagga population in the age of their imminent disappearance. The quagga were seen as having breeding potential. Attempts to tame and train them as labouring animals were only partially successful at scale. However, their indifference to the tsetse fly, and thus to specific illnesses that incapacitated domestic mammals in the work of empire in Africa, was valuable, and they had other value as food and for skins. In the 1820s Lord Morton bred a quagga stallion with an Arab mare in an attempt to breed back the quagga. He viewed the result as unsuccessful as the foal did not express quagga characteristics. However, a later foal from the same mare, when bred with a horse stallion, did express stripes on the legs. Morton wrote up his experiments for the Royal Society, and Enderbsy's account foregrounds the ways in which the concept of proof in the scientific method is shaped by the cultural specificity of the experiment. At the time, the results as Morton saw them were seen as supporting a theory of telegony, or the belief that offspring can inherit the characteristics of a previous mate of the female parent. Contemporary genetics holds telegony to be untrue and retrospectively explains this phenomenon as the result of dominant and recessive gene expression.

This story of how horses looked liked quaggas and quaggas looked liked horses is instructive in drawing out the mimetic importance of species characteristics, for the purposes of human definitions of species as a category. The quagga foal that looked like a horse we might now expect to have had a close genetic similarity to the quagga, although this wasn't expressed in terms of what it looked like, or what were thought of as the defining phenotypic quagga characteristics of colour and shape. The later foal of the horses, which we would now expect to have had more genetic similarity to horses, expressed visible characteristics that were associated with quaggas. However, in both cases sexual reproduction always had the capacity for surprising results. De-extinction, like cloning, is a kind of simulation of species, and the simulated dinosaurs of both *Jurassic Park* and nature documentaries, based on fossil evidence and natural history, have generated expectations about what extinct species look like. Illustrations and descriptions of woolly mammoths, together with specific fossil evidence of individual mammoths, have generated expectations about what mammoths are. Would an animal that looked like an Asian elephant but had a genome close to the mammoth's count as de-extinction? I suspect not. Any claims for the materialization of extinct species would have to meet visual as well as genetic expectations.

Just as the look, feel and texture of in vitro meat is part of the criteria for the reality of the substance as meat, so the mimesis of expectation is part of the criteria for the reality of genetically modified clones as forms of de-extinction. Thus, the space of flows that biomedia opens up allows the circulation of the representational and the material within the same form. Form and content matter; the biomedia and new fleshly realities of meat and species require a correlation of both, as the tissue of these animals becomes the medium for a new form of expression, and a space for mixing new media forms.

SPECIES HIERARCHY

Much contemporary cultural theory questions both race and species boundaries, through the turn to the animal in multiple fields, and anti-racist analysis in others. Writers such as Serres, Haraway, Wolfe, Shukin, Van Dooren, Ironside and many others explore multi-species worlds in which the microbial, viral and parasitic is constitutive of the host (Serres 1982; Ironstone 2011); in which flows can be traced across species barriers (Van Dooren 2014; Bird Rose 2012); and in which species meet and mingle (Haraway 2007). Practices in the contemporary sciences also deconstruct species at the same time as reconstructing a human centre. Work on the microbiome, for example, brings into question the idea that species exist in differentiated ways and opens up the horizon of entangled species assemblages. Cultural tropes such as viruses, contagion and parasites also facilitate thinking across and with species distinctions. While such tropes often operate in narratives that resolve in the sanctity of the human, they also open up new horizons of post-human possibility, and mass forms of popular culture are full of the undead (Ironstone 2011) and the nonhuman (Shukin 2009).

De-extinction, in its current form, reconstructs the human as an agent with the power to define, control and recreate species, and to undo climate change and habitat damage. It also reconstructs species as a contained and controllable category. It is, however, a very particular group of humans who are staking claims in this area. The groups of bioscientists involved in the meetings, projects, publications and discussions around de-extinction reproduce the structural inequalities of elite science and its institutions in the UK and the US. The people involved in this are predominately white men from elite institutions and businesses; there is some diversity in the teams but not much. These are not democratic or diverse projects. They are elite bioscience projects with a number of different agendas in play, from the role of George Church and Harvard, to that of *National Geographic* and the Beijing Genomics Institute. De-extinction works to bring genetic engineering, cloning and synthetic biology together under a hopeful sign which promises to restore the past, evoking nostalgic fantasies about a balance of nature, and a pastoral history of large-scale pre-industrial habitats. It provides something of a cover for advances in genetic engineering and synthetic biology which have seen opposition in the past. It enrols publics with the spectacle of charismatic species like the mammoth, through TED talks and popular science publications, while planning the creation of entirely new genetically engineered organisms that approximate the best illustrations that natural history has to offer.

FROM RECURSION TO INCURSION: MATERIAL AND UNREAL

Information, like humanity, cannot exist apart from the embodiment that brings it into being as a material entity in the world; and embodiment is always instantiated, local, and specific. Embodiment can be destroyed, but it cannot be replicated. Once the specific form constituting it is gone, no amount of massaging data will bring it back. This observation is as true of the planet as it is of an individual life-form. (Hayles 1999: 49)

Recursion, transduction and diffraction are powerful formulations to examine the dynamics of biomedia. These processes through which databases become points of corporealization are all ways of thinking about the movement of information and materials. Although in vitro meat is about deconstruction – uncowing the meat – and de-extinction is about simulation, they are both forms of material incursion. Much as biomedia are recursive, difference is made and these new forms are incursions into the real. New – as yet ontologically undefined – living materials have the potential to emerge in the world through in vitro meat and de-extinction. Genetically modified living forms and semi-living forms come into being and stabilize in moments of performative event, press releases, cookery shows, nature documentaries, live births. These are recursions of the visual and material texture of the thing promised, i.e. meat without the slaughter of the cow that looks like meat, and genetically modified cloned organisms that express characteristics of extinct species. They are recursions of the cells, DNA, tissues, sequences and information involved in the process of making – using forms of life to make life forms – but they are patterns diffracted, transduced and incursive in the real.

The language of biopolitics has come to signal a claustrophobic dead-end (Terranova 2009) in which life is made as capital (Shukin 2009; Cooper 2008; Mitchell and Waldby 2006; Sunder Rajan 2002). Transduction (Mackenzie 2002), diffraction (Haraway 1997; Barad 2014) and recursion (Jordan 2015; Kelty 2008), all offer more hopeful registers. Although life still might be the way that machines reproduce themselves (Bardini 2011; Dyer-Witheford 2015), this reproduction also offers the possibility of excess, suggesting that unpredictable differences are produced through the same means. As Haraway argues in relation to diffraction, Jordan in relation to recursion, and Mackenzie in relation to transduction, when the substrate or context changes, meaning and morphology also change. When the pattern of a genome is taken from a living organism and rendered as a sequence it changes meaning, shape and substance. When that pattern is used to make interventions in wetware things change again. When each difference is made the huge gulf between human understanding and intention on the one hand and the complexity of life and materials on the other is present in the process of making an intervention. In this gulf the potential for something beyond what is copied and known emerges.

De-extinction is a mode of reconstructing species, making new organisms in the look and shape of those now extinct. In vitro meat, on the other hand, is about deconstructing species, or uncowing the burger. It is also a form of simulation, reconstructing the texture and taste of meat. De-extinction and in vitro meat are simultaneously the most material and most unreal of the objects in this book. They are popular, I suggest, because of their capacity to put an object on the table. Like other technological fixes for the issues at hand they appear to offer more than models, processes and actions. They offer objects at the extreme. The material unreality of chimeric simulations which recursively prove their own reality through their materialization points to the power of the recursive to make incursions into materiality, to make up objects. This recursive generation offers a capacity for imaginaries to be rendered in the real; these are objects which offer an object orientated world, where things just are because they can be. The extraordinarily clear bridging of the object and text and the manipulation of what is real underscores the importance of taking objects as media forms. Under these conditions the political questions of whose ideas and whose appearances get made real becomes more urgent. These examples involve copying species in the name of genetic rescue and environmental restoration, or making more meat to solve the problems of meat production. They also facilitate the reproduction of the status quo of technocratic elites, creating highly developed technologies that promise much. These are closed systems of recursive representation that aim to reproduce their own values, narratives and imaginaries. Conversely, they are also closed systems made from unruly realities which will therefore produce unexpected elements, in excess of themselves. Whether these will generate enough escape velocity to get out of the cycle of objects offering a technological fix remains to be seen.

6

Unreal Objects and Political Realities

Genomes, biosensors and smart grids are objects that accelerate and intensify. They are ways of speeding up processes of mediation, technological change and capital. The genome speeds up and intensifies data creation, collection and processing. It speeds up the identification or diagnosis of diseases and other health conditions. But it doesn't decelerate, by which I mean it doesn't enable reflection or take account of experience, or of its own history. Smart grids speed up device production, fast-tracking the financing of some projects, but they inhibit reflection on a more collective approach. For example, the proposal to connect Iceland's energy resources to the UK mainland but to leave out Orkney is about grand promises and capital accumulation, not the pragmatics of national scales. The development of wind farms and solar panels out of materials at least as problematic as fossil fuels is not an inevitable technological given but a design choice. All these things can be changed but they have to be linked to collective political thinking, not driven by the technological object looking for a market. Accelerationism, in both its reach beyond the current horizon via technological evolutionism, and in iterations which head towards singularity, is on the side of the technological object looking for a market.

Each of the examples explored in this book can also offer forms of deceleration, of active subject making, of alternatives to the accelerationist and unreflective pace of emerging technologies.

Like accelerationism, and motivated in places by similar frustrations about texts and identity, new materialism and object orientation have become influential spaces of debate. New materialisms have operated as a rich nexus in feminist thinking, research and intervention in the last two decades. Feminist interventions have been centrally concerned with questions of material conditions and biological materialism. This is partly because they have struggled with embedded traditions about biology as destiny, and the classed, gendered, raced and sexual stratification of economic and other material conditions. Feminisms have, however, also been attentive to the role of the symbolic in shaping and reinforcing material conditions, and have intervened in politics that would have these separated. Donna Haraway's (1992) influential formulation of 'material-semiotic actors' cautions against the separation of things in themselves from their mediation. In some ways Haraway's intervention in the late 1980s and early 1990s was intended to cut against Bruno Latour's universe of actors in which things somehow stood outside of their mediation. She realigned science studies with work from gender studies which understood mediation as constitutive of things. Or, as other forms of feminism would have it, with bodies as the media of culture.

Haraway's intervention in the 1990s has strong parallels with feminist interventions in today's debates about objects, materials and mediation. At that earlier point, as innovative approaches in science studies emerged, particularly around actor-network theory (Latour 1991; Callon 1986, etc.), they appeared to ignore or side-track feminist history and epistemology. It took a long time for this area to enter into dialogue with or include work on how foundational categories such as sex, gender and race come to be made meaningful in the world. Nearly 30 years later there is another reinventing of the semiotic-material wheel and another failure to engage in dialogue with not just feminism but queer theory and critical race and disability studies. Object orientated philosophy and forms of speculative thinking such as accelerationism again seek to sever things from meaning-making practices and mediation in order to proclaim a flat ontology (Bogost 2012), a world of objects to be taken as such. Like actor-network theory 30 years ago, they appear to operate on a path isolated from many other fields. These discourses offer a flat universe of things, and have very little engagement with feminist interventions, which they seem to have forgotten or ignored. The citation practices in speculative realism, object orientated philosophy, accelerationism, and other areas such as critical theories of the digital are conspicuous in their disregard for feminisms. Their content also shares an impulse towards avoiding the difficulties of subjects and difference; eschewing the importance of meaning-making practices and mediation; elite, mystifying discourses; an epic scale that takes on both the molecular and the universe.

Perhaps more disheartening at this point is that there is not merely a side-stepping of feminist interventions in the mode of the 1990s, but the development of a post-feminist academic culture of its own. This is expressed in places by footnoting practice: Braidotti, Barad and Haraway get tacked on with little substantive treatment of their work. The current culture of critical theory is in places more redolent of post-feminist media cultures in which feminisms are deemed irrelevant because their goals are assumed to be already achieved. It is striking and symptomatic in this regard that object orientated approaches arose in games studies (Bogost 2006) simultaneously with the emergence of a violently misogynist media culture in gaming, evidenced by the gamergate controversy (Chess and Shaw 2015).

Object orientated approaches suggest a world that can be apprehended directly, given training in computation, or induction into the (right) language of theory (Bogost 2006; Morton 2013; Galloway et al. 2014). This could be an interesting constitutive contradiction – after all, code and language are rather semiotic – but this is side-stepped through the proposition that code and language are objects too. This is a form of purification where the messiness of texts, meaning making, semiotics, symbolism, fantasy and imagination are erased. Mediation and media fall away in the shadow of the object. However, such purifying moves don't change the world or resolve any of its problems, and these areas of debate become somewhat circular in both wheel-reinventing and citation practices.

OTHER INTERVENTIONS

The objects of technoscience are powerful phenomena in the world and they are highly political projects. That they are bracketed off from politics and offered as objects is part of the problem, and part of their seduction. Discussions about technology and culture have an attendant importance; they are part of the politics of a technoscientific culture which is inseparable from academia. Academia plays a role in the affirmation and mystification of objects, as well as sometimes enabling access to them. In each of the preceding chapters I have tried to trace out points of access, resistance or intervention.

Other contemporary discussions about technology and culture that do offer an opening might include the Xenofeminist Manifesto (Cubonicks 2015), Alexis Shotwell's *Against Purity* (2016), and Sarah Kember's *iMedia* (2015). Other projects that signal intervention include those hosted by Jenny Reardon's Science & Justice Research Center and her own writing about genomics and the political in *The Postgenomic*

Condition (2017). Preciado's (2013) work on the industrial production of sex, gender and reproductive forces also relocates technoscience as centrally structured through and by the regulation and control of gendered and sexed bodies, not just bodies or objects as given. These contributors all insist that identities, knowledges and worlds are forged through and in the processes of industrial-scale technoscience. Looking for the technological fix to take us out of here – whether designating the world as objects, or investing in more devices in the name of smart grids – creates more nodes of technoscientific entanglement. If we want to really get out of our present situation then collective engagement around the more irrational promises of monsters like in vitro meat, genome hacking, poetry for energy meters, or biosensor cat ears make just as much sense as investing in genomics while cutting basic health care.

Why unreal objects? I have used this term to excavate media materialities and science-media dynamics. This means looking at big technoscientific projects as media forms, not to deconstruct emerging technologies as hype, or promissory, but to evaluate them as constitutively media forms, not just as given objects. The genome is a medium through which both biological and cultural forms are materialized. However, embodied identities and their subjectivities are partly crafted through genomics – genomes offer individual and personal life stories, and identifications as well as structural categories like race, ethnicity and patient-hood.

Why unreal objects? Because it is important to continue to take account of the contingent, fragile and constantly negotiated state of knowledge in the sciences and the public understanding of technoscience more widely. The term 'science' remains indicative of a domain that secures knowledge making in the world, despite being subject to the same crises of knowledge elsewhere. Scientific meaning making across multiple kinds of technoscientific endeavour has become more insecure, subject to post-normal, post-truth economies in which mass attention on the one hand and exclusive invisibility on the other become more important determinants in legitimating actors in the world than questions of reality or truth. Public understanding of science becomes the domain in which scientific legitimacy is often made; however, this is not the kind of political utopia of science imagined in the move to open science or to democratizing science. It is a more complex and shifting terrain in which the stakes of knowledge production have become not just democratic but open to a wide field of publics including PR actors, investors, mediators, patient groups, innovators, artists, celebrity scientists and politicians.

In this context it is important to exercise care in approaching the question of knowledge production. Object orientated forms of material politics that focus on the world in terms of objects tend towards throwing out mediation altogether. For example, in *Hyperobjects*, Timothy Morton adopts an exemplary object orientated approach intended to persuade readers to accept that they are already always inside a series of objects by being in the world. This is presented as an appeal to ecological relationality and responsibility. However, it makes this move at the cost of abdicating responsibility for how ecology and responsibility are made meaningful. Like other versions of materialism, the story of hyperobjects reaches for scientific narratives – at least in their popular incarnations – to explain how knowledge can be directly made in the sensorium, no mediation required. Morton's analysis takes ideas about the selfish gene and the extended phenotype from Richard Dawkins' science writing on the same. This model of humanity as an expression, or side-effect, of the lively genome allows Morton to claim that we can know the world directly. For example: 'As I reach for the iPhone charger plugged into the dashboard, I reach into evolution, into the *extended phenotype* that doesn't stop at the edge of my skin but continues into all the spaces my humanness has colonized' (Morton 2013: 27). The problem with this version of the world as a layering of multidimensional objects is that it enables a world-view in which a mediator – Morton – provides insight into the world because it is made known directly through the writing voice. The extended phenotype that reaches into all that 'humanness has colonized' promises a direct biological, phenomenological experience of the world. This shuts out questions about how humans are also collectively responsible for colonization, for how the world is, as well as how it appears to be. Genomes and DNA become unassailable constructs through which a radical, interconnected world of hyperobjects is made known to readers through a writer who can articulate this direct experience of the world. Approaching the world as objects thus also precludes mediation and privileges particular ways of knowing in this mode. Morton's analysis cuts out all the work that has gone into making the genome meaningful such that it can be apprehended as an unquestionable object that structures phenomenology.

This approach to knowledge production through objects cuts into other ways of seeing. Galloway et al.'s *Excommunication* offers a much more media inclined but still object orientated approach to mediation, together with an equally epic world-view. The text offers up the uncanny and a sense of a wide vista beyond the known as new terrains for investigation. Not unlike *Hyperobjects*, it argues that what is important is knowledge production and experience beyond communication. It also expresses a frustration with the known and offers to move readers to desire inquiry beyond communication, representation and even life. However, it also makes this move at the expense of abdicating mediation, offering direct access to a process of knowledge production in which readers follow the authors into the beyond. Like the beyond in science fiction though, it can't yet be apprehended. The objects in the text of *Excommunication* are media objects, infused with liveliness, uncanniness, death and fury. But even as it enquires into mediation, the book also offers to abandon it.

The abdication of collective responsibility for meaning making in object orientated approaches is most explicit in Nick Srnicek's comment: 'Do we really need another analysis of how a cultural representation does symbolic violence to a marginal group? This is not to say that this work has been useless, just that it's become repetitive.' The comment is cited by Ian Bogost (2012: 132), who uses it as a platform to embrace a range of objects. Bogost is an influential games designer, scholar and critic, and some of his writing promotes and crafts speculative realism and object orientated approaches. However, in common with the discussions above, it is hard to see what orientation means in relation to the object. While the terminology of object orientation is drawn from programming languages, orientation is a situated disposition towards a thing which is embodied and meaningful. In her *Queer Phenomenology* (2006), Sarah Ahmed argues that objects are orientating devices, they order positioning and thus shape the meaning of both the orientated subject and the orientating object. To apply Ahmed's account of object orientation to Srnicek, Bogost and Morton's objects would be to displace their privileging of the object back onto the question of how an orientation puts some things in reach. In the discussions about speculation and objects that I have traced, there is an abdication of mediation, representation, communication and meaning making in the making up of objects. This means that we are left in a world without subjects, in which unmediated access to objects (especially technoscientific ones) is set up as the way of knowing. This politics of knowledge production orientates particular bodies, identities and subjects as reaching for those objects and cuts out others.

It specifically cuts out feminist media scholars, critical race scholars and other intersectional, situated knowledges. It cuts out the repetitive work of coming back to questions of representation and violence, which are now more necessary than ever.

Object oriented discussions privilege specific objects, articulate frustration with meaning making and repeated analyses of the relationship between representation and violence, and occlude situation, subjectivity and identities. Such frustrations and occlusions also feature in technoscientific research and innovation contexts. In response to them it is tempting to think about moving away from elite objects like genomes and smart grids. But, thinking with Donna Haraway, it is also worth persevering with the trouble that these objects bring. This book then tries to reorientate the process of constituting objects, which means taking seriously the challenge of thinking about objects and subjects as mediated. This is also the challenge of thinking about technoscientific objects as both things which appear to be in the world and as media forms in which identities, knowledges and worlds are made and can be remade.

It is an approach that brings materialist orientations and situated positions together, treating genomes, biosensors, smart grids, tissue-engineered entities and clones as objects but also drawing out all the work that goes into making them appear so. They are both actual things in the world that orientate people, and at the same time media forms, or generative meaning-making practices. Where Haraway's work stood in relation to actor-network theory in the 1990s is in parallel with feminist interventions in relation to object materialism in the current moment. This parallel situates my own work in relation to debates loosely centred around the term 'new materialism' in feminist engagements with technology and culture. However, I also bring to this a strong sense of mediation as central, drawing on Kember and Zylinska's work on mediation as a vital process and tying this in with the interventions of Cubonicks (2015), Power (2015, 2016) and Asberg et al. (2015).

MEDIA MATERIALITIES

This book is in part an attempt to create accountability by disorientating objects. With this in mind the chapters have explored emerging sciences and technologies at multiple sites in which they are made real. The mass-market roll out or framing of the object in each case has been examined to suggest a range of alternatives for looking at how things might be otherwise. This is intended to open up these sites of technoscience in the making in order that something like accountability, or at least accessibility, might be facilitated.

Ian Welsh and Brian Wynne (2013) also argue for accountability in relation to technoscience. They have argued that there is a culture of scientism in the UK in which science operates as a surrogate politics. Consequently, they call for accountability in elite discourses around science and technology. They advocate this as a way of making science and politics more robust and democratic: 'This kind of critical work in no way diminishes science; rather, it calls for an explicit deliberative politics around normative questions hidden in adjudications that are packaged as scientific or technological' (2013: 562). That might mean disrupting the seamlessness of elite discourses where normative political judgements are already encoded in scientific projects; for example, the taken-for-granted ideas that a national genomics project is important or that fitness tracking and smart energy metering are necessarily good things. It means disrupting the construction of technoscientific objects as given.

Maureen McNeil (2013) argues that scholarship in science and technology studies has been part of the problem in maintaining science as an elite discourse. Models of both science and publics in the field have limited the capacity for more democratic modes of science, partly because of a reliance on a diffusion model of science in which laboratories are its 'fact factories' (McNeil 2013: 601). I have brought these debates in science and technology studies together with debates in media studies, particularly those around materialism, digital media and information politics. My analysis traces out how theoretical writing can obfuscate the project of making science and technology more accountable. For example, some versions of the turn to materialism represent science and technology as the most important or indeed only site of intervention, or the only lens through which the world can be apprehended (Bogost 2012; Morton 2013). Other versions claim that we already know enough about culture and politics and must look beyond these to something unreachable with available modes of criticism (Galloway et al. 2014). Still other diagnoses posit that the problem is that we are not sufficiently attuned to the physical and material world (Bennett 2009); while some commentators valorize the discourses of science and technology at the expense of other voices and stories (Grosz 2011). These perspectives compound visions of technoscientific objects as the central, most powerful force, and in doing so make it less clear how participation and, most importantly, the possibility of intervention could be realized.

Academic debates have turned to materiality at the same time as the world around us has seen a rise in unreal objects. In his book *Material Politics*, Andrew Barry makes an observation that emphasizes this point: 'it is ironic however, that just as social theorists and philosophers are increasingly drawing our attention to the agency of materials, the properties and activities of materials have progressively become the objects of increasing levels of information production' (2013: 13). Materiality is so thoroughly interwoven with information, imaginaries and different conditions of contested meaning that it is even more important to develop frameworks that examine differential and shifting materialities and their mediation. To pursue this, the preceding chapters have explored several unreal object case studies together with a discussion of academic debates about materialism in the fields of media and communication studies and science and technology studies. The underlying political question is about the constitution of things that matter (cf. Butler 1993): how and why some things come into the orbit of attention, investment and care, while others are shut out. There is an overwhelming attachment to technoscientific things and realities at the expense of others, and the preceding chapters have been attempts to tell stories about these unreal objects and offer some alternatives.

Unreal objects can't just be wished away; they are part of the making of political realities. Objects invoke public gathering, operating as the matters of concern around which politics are made (Latour and Weibel 2005). However, matters of concern are made in media forms through which those issues are instantiated. What comes to matter as a political centre is often the mediation of an imaginary object, or an object only held together as such through laboured representation. We attend to the concerns written into these, but instead of recognizing them as fictions, imaginaries and speculations, we take them as stand-ins for matters of fact. When we look again at stories in the contemporary moment we have to recognize that such materialities inhere in media forms. They register patterns in light which register on retinas, provoke phenomenological experiences, generate affect, make meaning. They can be touched, seen, heard and enacted. They have an emotional charge and intellectual resonance. Forms previously understood as symbolic and representational (and even ephemeral) manifest across a range of material registers.

The category of the real as actual, authentic or existing applies to fact and fiction, is felt, thought, seen, read and experienced. To talk of unreal objects is to highlight an expansion of the real beyond objects, to invite a recognition of the power of things seen as less real, and to encourage storytelling as a form of intervention.

MATERIALIST IDENTITY POLITICS

All of the examples in this book have different qualities in relation to the terms material, real and object. However, they all appear as given objects in the world. The gathering of particular elites around them demands a return to questions of identity politics, but perhaps with a more expansive inflection. How and why are particular technocratic elites reproducing their own power and visions of the world? The turn to materialism in academic debates is in part motivated by a reaction against and refusal of questions of representation and identity politics. However, when particular elites are in control of economic, media and biological production, questions about whose worlds are in becoming, and who is doing the world-making, are important, and require analyses of identity politics and representation as well as the material and economic aspects.

Genomes are representational and relational texts but they are taken as a kind of informatic real, given as objects. In media and communication studies questions of representation, seen as dominating the field at one point, have been cast aside in some quarters. The argument that objects need to come to the fore only works by evacuating a history of media materialism from a history of media studies. This enables a strong and somewhat inaccurate claim for a new framework of analysis to be launched (Fred Turner's (2014) work is useful in demonstrating this point further). Ghosts of materialism and technology in the history of media studies, such as Raymond Williams and Marshall McLuhan, have been reclaimed, sometimes as newly found, and translations of and responses to Kittler (Parikka 2013) and Flusser have become newly foundational. However, although material approaches are vital they don't have to be pursued at the expense of mediation. The more that informatic forms of representation such as data are taken as worldly materials the more important it is to foreground this.

Some of these academic debates about materialism are outward looking and propelled by a concern that the rise of immaterial interfaces (e.g. social media) intensifies a technicist ideology. There is a concern that digital media in particular conceal their own means of production, their material and artefactual elements, such as code layers and environmental damage. Bringing together science and technology studies and media studies can be a fruitful move in addressing these concerns, and Gillespie et al.'s (2014) collection about media technologies is useful in doing this. As already noted, there is also a rich literature in feminist materialism which registers high visibility with publications and conferences multiplying. Figures whose work directly engages with this field include Coole and Frost (2010), Van der Tuin and Dolphijn (2012), Colebrook (2008) and Alaimo and Heckman (2008). Although Colebrook and Van der Tuin associate feminist new materialism with third wave feminism their work also traces associations through Haraway, Braidotti, Kirby and Grosz. Irigaray and Simone de Beauvoir often operate as key foundational figures in this area, as in other feminist trajectories.

This book has charted a path through these theoretical debates, while examining some real world objects, texts and practices. In tune with Haraway's version of the material-semiotic (1992), and Kember and Zylinska's (2012) call to look at media realities after new media, this is a project about the politics of unreal objects. It has demonstrated differentials in materialities as a strategy to reframe and subvert material-immaterial distinctions. In doing so it has shown the capacity for objects that are largely unreal, whatever their materiality, to capture both the political imaginary and economic investment; and it contextualizes genomics, biosensors, smart grids and biomedia through this framework.

TECHNOSCIENTIFIC OBJECTS AND MEDIA LIFE

Genomes make life as media, promising that the same reading and editing processes that shape the composing of documents or code also shape the composition of the human. The circulation of genomes reinforces Mark Deuze's (2012) argument that life is lived in media, but can also be opened up to Kember and Zylinska's suggestion that we 'shift from thinking about media solely as things at our disposal to recognizing our entanglement with media at a sociocultural as well as biological level' (2012: 1). Kember and Zylinksa argue that mediation as a dynamic process can be thought of in terms of flows of being in the world, vital processes, in which media artefacts are stabilizations or cuts. Genomes are a specific stabilization of media and biological life, and our entanglement with them illustrates how understandings of life itself are bound up in genomes both as objects and as part of a process of mediation.

Biosensors are unreal objects because while they are material electronic devices, they anchor an unreal imaginary of joined-up health-care systems and big data. This imaginary is instituted in the objects themselves, the media representations and discourses about them, and the interfaces with the objects and the data they collect. While people do carry around these objects and data is produced in the form of measurements of bodily signs, there is no system of joined-up health care or even big data that these sensors belong to. They generate and contribute to individual databases and changes in behaviour, and are effective forms of surveillance. However, the fantasy of empowered pre-emptive patient-consumers who will regulate their own health in relation to normative disciplinary discourses, while freeing up medical resources and improving the general health of the population, remains unrealized. Likewise, the idea that the collection of enough data will generate new insights into health and medicine remains aspirational.

Biosensors are distractingly material at the level of the individual device to the point that they obscure the bigger scale at which they are unreal and damaging. Taken out of their preferred designation, they can be seen to have unexpected connections with older phenomena. For example, as we have seen, a device like Fitbit can be thought of in a genealogy of technologies disciplining women, like the daily letters and diaries of nineteenth-century middle-class women of the colonial structure.

Biosensors intersect with critical discourses of materialism because they appear to offer a way of understanding the world beyond the human sensorium. They deploy a range of sense perceptions that are not directly or easily available, from counting steps and monitoring sleep to the promise of detecting brainwaves. They offer a glimpse into what it might mean to use multiple forms of sense perception that are outside of the human range. This is a radical promise that opens up the possibility that people might enter into communication with the world in new ways. This promise is explored in the biosensor avant-garde in art work and hacker interventions; for example, in biosensor projects that make visible the interactions of microbes and environment. These impulses to find new ways of knowing intersect with the aspirations of forms of materialism because they offer a new informatic way of apprehending the world. They promise a possible answer to the vexed question of how we know about the world around us, if not through media representations. If materialism has given up on media and interpretation then biosensors seem to offer a new sensory and informatic paradigm that gives us a different vision of the world. But this tentative promise is illusory. Biosensors provide another layer of representation that is produced, consumed and circulated through the prism of mediation and interpretation. However radical the promise that people might access new forms of perception, these are always recuperated into human language, and biosensors remain media technologies.

Smart grids promise new systems of energy control and regulation that scale up to the planetary. They also look to de-centre the human and offer visions of informatic energy futures. They have materialized in the form of an industry and media materials, from policy documents, proposals and forward-looking statements to advertising and descriptions of systems. Elements of the smart grid imaginary have also been developed: smart meters that measure power use; and components of sustainable fuel sources such as technologies for solar and wind capture. However, the vision of this as a new and world-changing energy system that will use big data to engineer sustainable, carbon-neutral energy futures is not close to materialization. In the meantime the redirection of public and private money into investments in energy futures, and the materialization of another generation of electronic devices, is an extension of the current disastrous energy economy. Thus, while smart grids reach for global joined-up power systems and promise a more sustainable energy future, in the here and now they contribute to the problems of unsustainable power and technology use.

Smart grids are a powerful imaginary, producing a total vision of new energy futures with the flexibility to incorporate old ones. However, much of what is said about them is entirely fictitious and speculative, and thus very open to intervention and reworking. Although decision-making in this area is masked by stories about inevitable objects, the smart meter is embedded in existing infrastructural path dependencies, not in new smart grids (Bowker et al. 2010). This raises the question of which speculations and fictions get to count as real futures. This is interesting to think about in the context of global warming, which has itself been treated as fictional or only a model (Edwards 2010). The warnings of climate change scientists, activists and lobbyists were set aside as fictions for decades. Although they have gained traction as factual forms in the last 15 years, this remains precarious, and investment in fracking, together with US policy on the issue, impedes the stabilization of any scientific consensus. However, the smart grid imaginary has not been controversial or called out as fiction because it appears as a fixing object. In the world of emerging technologies and energy policy some fictions get designated as such and pushed out of political consideration, while those that attach to objects get taken with great seriousness and attract investment.

Computing power appears in the smart grid imaginary as something newly added to the electrical system, although conversely computing can also be seen as electricity made visible. In grid terms it is also imagined as a utility, like electricity. These imaginaries invoke political utopias, alliances across antagonisms, and reversals of older power relations. If these could be realized the political transformations could be inspiring, but it is worth looking back to the vision of Atlantropa to think about the multiple factors and lived details that are missed in epic engineering dreams.

De-extinction and in vitro meat offer a different way of thinking about unreal objects, and a useful counterpoint to the previous examples. The latter all offer to make the world as informatic objects: genomes, data, devices, networks. The examples in Chapter 5 could be taken as what happens when those informatic forms are used as the basis for rendering bodies and worlds. If the informatic mode takes the complexity of lived experiences, human bodies and whole ecosystems and renders them digital, then things are changed in those processes. The recursive, transductive and diffractive dynamics of making things in an informatic mode make things differently material. These examples are about creating new bodies, or forms of embodiment, from informatic materials. They are new incursions into the real and they signal what things might look like if the hubris of technological control is taken as truth.

These interweavings of real/unreal, material, media and ontology look like a cat's cradle of epic proportions, hyperobjects of our own making. The material effects of unreal objects, defined as realities through the discourses of science policy and technology innovation, are devastating. Without a full acknowledgement of the extent of fiction and fantasy in everyday life, science, technology and policy, we are in danger of creating conditions of such complexity and confusion that it becomes impossible to act. The discourses of complexity, of big data and of hyperobjects create ideological folds within which the world is conceived as something of such complexity that only very elite and specialized fields of expertise can be brought to bear on knowledge making. Although the unfolding of the worlds in which we live is beyond individual human perception, many of the most pressing issues of our time are also to do with the obscuring of problems that could be addressed if we could cut through mystification and epic claims to complexity beyond comprehension. A collective, rather than elite, address to current issues would open them up to multiple viewpoints, and this multiplicity in itself offers a richer promise of a view from beyond individual human perception than that of machine vision.

DATA OBJECTS AS DISTANCE

All of the examples in this book are caught up in data production and representation. Genomics makes bodies, flesh, identities and species known through the lens of sequence data. Biosensors collect biological signals across a range of variables, including movement, sweat, blood sugar, sleep and heat, and put these together in interfaces and representations that become ways of knowing bodies. Smart grids represent the world of humans, consumption and energy as a network of data points and processes that can be modulated and controlled. De-extinction and in vitro meat are ways of taking data representations of bodies and using them as the basis for creating new organisms.

The forms of scientism embedded in the current expansion of data into all areas of life are not only contested in the arts and humanities but also taken up in those sites. Franco Moretti, for example, is well known for employing data analysis and data representation in the service of new methods in the humanities. Moretti's work on distant reading proposes big data as a method for understanding literature. He argues that scholarship cannot understand a particular literary period through close reading because the sample size is just too small. Instead he proposes a model of distant reading in which data representations of a literary period are taken instead. In Moretti's model of analysis the unit of data collection, in his case the novel, becomes less meaningful and disappears as relevant even as it is subject to analysis:

> Distant reading: where distance, let me repeat, is a condition of knowledge: it allows you to focus on units that are much smaller or much larger than the text. Devices, themes, tropes – or genres and systems. And if between the very small and the very large the text

itself disappears, well it is one of those cases where we can justifiably say, less is more. If we want to understand the system in its entirety we must accept losing something. (Moretti 2000: 48)

This sounds a lot like the distant reading of data representations writ large, whether population data, genomic data or social media analytics. On the one hand this approach offers the promise of understanding something in its entirety or as a whole system. At the same time something is lost, the unit of analysis that has been aggregated as an object disappears from view. In the case of genomes, the genome becomes irrelevant except as a collection of 100,000 genomes. The organism from which the pattern is derived disappears from view and the promise of new drugs or newly modified genomes comes into view. The organism disappears to be replaced by aggregate genomes. For example, in the proposal to write a synthetic genome, the living organism that the genome is supposed to be a means of care for disappears entirely. What then happens if this new understanding is used to simulate the original unit of analysis, and what if the units that disappear from view are not just books, but people and things?

This seems very close to the flat ontology of object orientations like those of Bogost, Bryant and Morton: 'An ontology is flat if it makes no distinction between the types of things that exist but treats all equally' (Bogost 2012: 17). In this proposition it appears that nothing is lost if things are just flattened out. It does, however, involve another kind of loss, that of specificity and attention to difference. This idea makes very little sense in the context of worlds which are shaped by the distinctions between types of things: animals, capital, countries, disciplines. When whole systems of governance, politics, identity and life are lived out in relation to the meanings attached to distinctions between things, the question is still, and to repeat, surely more about what and how things come to matter (Barad 2007; Butler 1993).

Object materialisms propose a democracy of things, flat ontologies in which meaningful distinctions can't be made between different kinds of being. This enclosure leads to another: they are entirely singular in which practices and approaches they advocate. This produces exceptionalist methods; for example, Bogost is not for a plurality of approaches, and Morton doesn't allow that representation really matters.

In contradistinction to these single-track approaches, it might be useful to take Isabelle Stengers as offering the final word on materialism:

The challenge, which I deem a materialist challenge, is that whatever the mess and perplexity that may result, we should resist the temptation to pick and choose among practices – keeping those which appear rational and judging away the others, tarot-card reading, for instance. (2011: 379)

This is a useful proposition to think with about how to collect objects. In vitro meat and de-extinction might be the tarot-card readings, while genomics and biosensors sometimes appear rational. In relation to genomics it might be tempting to put the PR work and gnomes alongside the tarot reading, and the sequencing of patient genomes in the rational category. However, they are both elements of the object of genomics. The preceding chapters on biosensors and smart grids are characterized by a collective approach to both tarot reading and things that appear more rational, in which Necomimi cat ears, Fitbits, novels and dreams can be thought of as parts of a whole collection of practices. If the chapter offers a flat ontology in which the cat ears and Fitbit are equally the object of biosensors, it also brings in mediation and orientation. Fitbit offers to make rational subjects, while the cat ears might offer a more wild-card orientation.

Some of this book has been taken up with contesting versions of object orientated and speculative materialist theory for its epic reach, lack of situatedness, glamorization of science and technology, and disavowal of representation. I contested these propositions because it feels like they make the question of how to intervene in the world less accessible. They seem to offer up objects as given and in relation to which the only subject position is reaction. Feminist approaches to materialism, on the other hand, have used the materiality of sex and race, and conditions of labour, wealth and inequality, as grounds for intervention. In the face of conditions in which science and technology become surrogate politics, authorizing legitimacy, and where visions of the world as complex and doomed, or as salvageable only through more science and technology, persist in both technoscience and critical theory, the grounds for intervention become difficult to see. However, if things are recognized as manufactured in the work it takes to story them in the world, then those grounds become more open.

Unreal objects are made up though mixtures of real and unreal, fiction and fact, stories and things. The issue becomes one of recognizing them as such and turning away from the fetishization of emerging technologies as definitive object realities. Such an orientation might indicate that being open to the reality of fictions allows a creative ground for intervention. We might recognize genomics as a fictitious luxury project enriching the shareholders of sequencing companies, and thus decide to put more money into basic health care instead. Biosensors could be recalibrated and retuned to tell us something new instead of counting our steps. We might re-story smart grids as a collective gathering around the problem of sourcing renewable energy and take those nodes that already exist as ways forward, instead of inventing an industry of smart meters and a pipeline to Iceland. We might look at the story arcs of de-extinction and in vitro meat and conclude that if the promises made in these stories are so compelling, then we might explore a more imaginative, less elitist and more collective route to the same narrative end-point.

The point then might be to take the world as 'fat and living' (Love 2010: 381), and to think about a politics of attachment rather than distance. To offer ways into the elite discourses of technoscience, rather than mystifying them further, and to consider what a feminist approach to data and materials might be. A feminist lens on data is useful in deconstructing its apparent neutrality and objectivity and examining the way it is cooked up (Gitleman 2013; Kennedy 2015). The informational representations offered through data production mechanisms like genomics, biosensors and smart grids are of course as carefully constructed as any other media text. They have their own framings, selection processes, rhetorical devices and visual culture. In the mode of counting we have to remember to count ourselves back in, not as units of measurement but as the embodied, messy experiential forms of living from which data is extracted. Each preceding chapter's foray into different modes of storytelling about the objects concerned has been an attempt to do this. Hence, art projects, histories and genealogies, counter-discourses and experiences have been set alongside the more dominant shadows of these objects.

In the case of genomics, the bodies that matter are those from which genomes are sourced, and this works both ways. Elite bodies get sequenced first, and the patients of value to sequence collectors get some forms of care. Genomes are quickly removed from those bodies though, and reproduced as scaled-up beautiful patterns of promise. At that remove they lose connection to the living and the fat and become an ephemeral promise which it is difficult to gather around in meaningful ways. The question of who they benefit and what power they have becomes assumed as part of scientific decision-making, when these questions have already become surrogate politics, allowing the flow of public money to private beneficiaries.

Biosensors in their fitness tracking mode become new forms of disciplining women through normative visions of better, fitter lives. But they also open up the promise of coming to know the world differently. While they manifest as indefinite streams of data colonizing, settling in interfaces that tell constricted stories about health and fitness, alternative ways of putting people, bodies and artefacts back into this have been pursued through art practice and some research and development. However, the promise of a critical biosensing that would open up the world in new ways is always frustrated by the recapitulation of existing formats for data visualization and its relationship to language. That is to say, new forms are expressed through conventional formats. New stories about the relationship between self, environment and device help to unfold these devices into processes of mediation and put bodies back in. However, this takes work to realize and doesn't come from biosensors as they are currently offered but through playing with them and breaking them up, putting them in alignment with other media forms and different forms of storytelling.

POLITICAL REALITIES

Any claim to speak of political realities needs to be grounded in questions about whose politics, or whose realities, are involved. I am writing this conclusion as a UK national in a particular political landscape in which austerity, violence, racism and sexism have been heightened, and after the UK voted to leave the EU. I suggested in the introduction that unreal objects are so appealing precisely because the political conditions and experiences of so many groups of disenfranchised people seem so appalling. The technological horizon becomes an orientation towards an escape route from the present and the real. One current figure for this is the vision of space travel to Mars, where the object of the spaceship erases a possible gathering around the politics of climate change.

This book was partly informed by my participation in an EU-funded project that examined the social aspects of emerging technologies and tried to make an intervention in questions of technology assessment. Currently there is uncertainty among those who have until recently been counted as Europeans about who is in Europe, and there is also uncertainty among those who have been counted as in the UK as to what that means. Identity politics remain important, and the sense of a political, civic and national self is also central to decision-making, allocating resources and gathering around issues and objects. Genomics England promises benefits for the UK as a whole (despite the bracketing off of England in the title), and states on its website that it will result in 'a country which hosts the world's leading genomic companies'. At the same time, the meaning of the UK itself, and of what it means to be a country, are also made in this claim to have a national genomics project.

One political reality for almost everyone is that the media is at the heart of contemporary technoculture, but that this centrality is almost unseen. The ubiquity of mediatization ensures its invisibility and its very centrality leads to its disavowal. In the UK political and media spheres, media studies are denigrated, dismissed as problematic, reductive, or transparent and simple. Likewise, media professionals get positioned as sensationalist, corrupt, sensation seeking, superficial. Popular and mass media forms are positioned as at best superficial and at worst sites of brainwashing, distractive forms of passive consumption. It is true that media forms are not well regulated or always pursuing an ethical relation, and that journalism is experiencing an ongoing crisis, but it is also true that, despite their ubiquity, they are not well understood.

Mediation is a vital process in a mediated society. The rise of media materials and the ascendance of public relations and communications in all industries shapes knowledge making and legitimation. Public relations professionals and agencies like the Thin Air Factory or the Department of Expansion have promoted genomics and in vitro meat at an early stage. Media relations centres such as the Science Media Centre, or actors like the Wellcome Trust who have the funds to help generate media cultures promoting the biosciences, are all significant players in shaping the kinds of science and technology that political forces gather around. Hollywood films, animations, art, social media, television drama, news and documentary all play a role in making scientific realities. PR animations designed to engage publics around the meaning of genomics distract from questions of political economy, such as whether public funds should go to private sequencing companies.

The question of which visions get to count as political realities becomes linked to who can afford the best media production values. This in turn links to the question of who has the most impressive object to put on the table. The convergence of science, technology and media creates the conditions for unreal dreamscapes to become political realities. They become realities because they take up resources, and exacerbate existing problems. They take up time, thought and attention that might otherwise be placed elsewhere, and they have effects on people's lives.

UNREAL FUTURES

Given the contradiction of a technoscientific society in which emerging technologies are both positioned as the fix to all problems and contribute to creating those same problems, how could those technologies be different? If emerging technologies come to the table as already formed objects, how could they be made up otherwise? This book demonstrates the central role of the media in object making and in shaping the reality of emerging technoscience. It also opens up the question of how to make up objects differently.

Activist and artistic engagement and making practices, and creative interstices in the popular culture of science, are places where differently placed fictions, illusions, visions and imaginaries are already part of the making of technological futures. Taking those registers which don't disavow their own fictions but foreground them allows us to look at the power of making things up in a more positive way.

Admitting first and foremost at policy, governance and education levels that emerging technologies are also fictions that have a role in making up the world would enable a more open engagement with the question of where public money, attention and education resources might be placed. This kind of framing is already dominant in some design practices, such as design fictions and participatory design (Dunne and Raby 2013), and in future ethnography (Watts et al. 2014; Watts 2016). A more open discussion about the power of mediated technoscience as the central dream factory of technocultures would enable a demystification of these areas and open them up to a wider engagement, perhaps enabling them to become attuned to more sustainable, more equitable and more creative futures.

Making things up collectively, and appreciating our collective investments in fictions as we go along, is a better bet than investing in engineer's dreams such as draining the Mediterranean or burying carbon under the sea. Creative responses to the challenges of the present moment are necessary to get us out of here. A greater collective engagement with imagining futures, and a collective project to address injustices, are clearly possible. Technoscience has an extraordinary capacity to put objects on the table, but those objects don't have to be taken as the only realities available.

Bibliography

Ahmed, S. (2006) *Queer Phenomenology: Orientations, Objects, Others*. Duke University Press.

Ahmed, S. (2010) *The Promise of Happiness*. Duke University Press.

Alaimo, S. and Heckman, S. (eds) (2008) *Material Feminisms*. Indiana University Press.

Allan, S., Adam, A. and Carter, C. (eds) (2000) *Environmental Risks and the Media*. Routledge.

Allison, A. (2003) 'Portable Monsters and Commodity Cuteness: Pokémon as Japan's New Global Power', *Postcolonial Studies* 6(3): 381–95.

Anderson, E. (2015) 'Tour of the Monuments of Silicon Valley', UCSC, http://artsites.ucsc.edu/faculty/eanderson/silicon/video.html.

Asberg, C., Thiele, K. and Van der Tuin, I. (2015) 'Speculative Before the Turn: Reintroducing Feminist Materialist Performativity', *Cultural Studies Review* 21(2): 145–72.

Atkinson, P. (2015) 'Thinking With Digits: Cinema and the Digital-Analogue Opposition', *Sequence* 4.1 (unpaginated).

Bainbridge, J. (2013) '"It is a Pokémon World": The Pokémon Franchise and the Environment', *International Journal of Cultural Studies*, 17(4): 399–414.

Baldwin, N. C. (1960) *Fifty Years of British Air Mails 1911–1960*. Field.

Bar, F. and Galperin, H. (2005) 'Geeks, Bureaucrats and Cowboys: Deploying Internet Infrastructure, the Wireless Way', in Castells and Cardoso 2005.

Barad, K. (2003) 'Posthumanist Performativity: Toward an Understanding of How Matter Comes to Matter', *Signs: Journal of Women in Culture and Society* 28(3): 801–31.

Barad, K. (2007) *Meeting the Universe Halfway: Quantum Physics and the Entanglement of Matter*. Duke University Press.

Barad, K. (2014) 'Diffracting Diffraction: Cutting Together-Apart', *Parallax* 20(3): 168–87.

Bardini, T. (2011) *Junkware*. University of Minnesota Press.

Barry, A. (2013) *Material Politics: Disputes Along the Pipeline*. Wiley.

Bassett, C. (1999) 'A Manifesto Against Manifestos', in Sollfrank, C. and Old Boys Network (eds), *Next Cyberfeminist International*. Old Boys Network.

Bassett, C. (2015) 'Feminism, Expertise and the Computational Turn', in Thornham, H. and Weissmann, E. (eds), *Renewing Feminism: Radical Narratives, Fantasies and Futures in Media Studies*. I.B. Tauris.

Batchen, G. (2006) 'Electricity Made Visible', in Chun, W. and Keenan, T. (eds), *New Media, Old Media: A History and Theory Reader*. Routledge.

Battaglia, D. (2001) 'Multiplicities: An Anthropologist's Thoughts on Replicants and Clones in Popular Film', *Critical Inquiry* 27(3): 493–514.

Beaulieu, A., de Wilde, J. and Scherpen, J. (eds) (2016) *Smart Grids from a Global Perspective: Bridging Old and New Energy Systems*. Springer.

Bell, D. (1973) *The Coming of the Post-Industrial Society: A Venture in Social Forecasting*. Harper.

Bennett, J. (2009) *Vibrant Matter*. Duke University Press.

Benson-Allott, C. (2015) 'Editor's Introduction', *Feminist Media Histories*, 1(3): 1–3.

Berry, D. (2014) *Critical Theory and the Digital*. Bloomsbury Press.

Bird Rose, D. (2012) 'Multispecies Knots of Ethical Time', *Environmental Philosophy* 9(1): 127–40.

Bogost, I. (2006) *Unit Operations: An Approach to Video Game Criticism*. MIT Press.

Bogost, I. (2012) *Alien Phenomenology, Or, What It's Like to be a Thing*. University of Minnesota Press.

Bolter, J. and Grusin, R. (1998) *Remediation: Understanding New Media*. MIT Press.

Botting, F. (1999) 'Virtual Romanticism', in Larrisy, E. (ed.), *Romanticism and Postmodernism*. Cambridge University Press.

boyd, D. and Crawford, K. (2012) 'Critical Questions For Big Data: Provocations for a Cultural, Technological, and Scholarly Phenomenon', *Information, Communication and Society* 15(5): 662–79.

Boykoff, M. and Boykoff, J. M. (2004) 'Balance as Bias: Global Warming and the US Prestige Press', *Global Environmental Change* 14: 125–13.

Bowker, G., Baker, K., Millerand, F. and Ribes, D. (2010) 'Toward Information Infrastructure Studies: Ways of Knowing in a Networked Environment', in Hunsinger, J. (ed.), *International Handbook of Internet Research*. Springer.

Braidotti, R. (2001) *Metamorphoses: Towards a Materialist Theory of Becoming*. Polity.

Brennan, E. (2013) '"Debate is Idiot Distraction": Accelerationism and the Politics of the Internet', *3: AM Magazine*, www.3ammagazine.com/3am/debate-is-idiot-distraction-accelerationism-and-the-politics-of-the-internet.

Brown, N., Kraft, A. and Martin, P. (2006) 'The Promissory Pasts of Blood Stem Cells', *Biosocieties* 1(3): 329–48.

Buchanan, K., Russo, R. and Anderson, B. (2015) 'The Question of Energy Reduction: The Problem(s) with Feedback', *Energy Policy* 77: 89–96.

Butler, J. (1993) *Bodies that Matter: On the Discursive Limits of 'Sex'*. Routledge.

Callon, M. (1986) 'Elements of a Sociology of Translation: Domestication of the Scallops and the Fishermen of St Brieuc Bay', in Law, J. (ed.), *Power, Action and Belief: A New Sociology of Knowledge*. Routledge.

Campbell-Smith, D. (2011) *Masters of the Post: The Authorized History of the Royal Mail*. Penguin.

Carey, J. W. (2009) *A Cultural Approach to Communication: Communication as Culture*. Routledge.

Cartwright, L. (1995) *Screening the Body: Tracing Medicine's Visual Culture*. University of Minnesota Press.

Castells, M. (1989) *The Informational City: Information Technology, Economic Restructuring, and the Urban-Regional Process*. Blackwell.

Castells, M. (1996) *The Rise of the Network Society. The Information Age: Economy, Society and Culture*, Vol. 1. Blackwell.

Castells, M. (2012) *Networks of Outrage and Hope: Social Movements in the Internet Age*. Polity.

Castells, M. and Cardoso, G. (eds) (2005) *The Network Society: From Knowledge to Policy*. Johns Hopkins Center for Transatlantic Relations.

Catts, O. (undated) 'The Art of the Semi-Living', www.tca.uwa.edu.au.

Catts, O. and Zurr, I. (2005) 'Big Pigs, Small Wings: On Genohype and Artistic Autonomy', *Culture Machine*, 7 (n.p.).

Cheney-Lippold, J. (2011) 'A New Algorithmic Identity: Soft Biopolitics and the Modulation of Control', *Theory, Culture & Society* 28: 164–81.

Chess, S. and Shaw, A. (2015) 'A Conspiracy of Fishes, or, How We Learned to Stop Worrying About #GamerGate and Embrace Hegemonic Masculinity', *Journal of Broadcasting and Electronic Media* 59(1): 208–20.

Christie, A. (1939) *Murder is Easy*. William Collins and Sons.

Christie, A. (1957) *4.50 From Paddington*. William Collins and Sons.

Chun, W. (2011) 'Crisis, Crisis, Crisis, or Sovereignty and Networks', *Theory, Culture & Society* 28(6): 91–112.

Cipriani, J. (2015) 'Here's Why Fitbit is Giving Target 335,000 Fitness-tracking Devices', *Fortune*, 16 September.

Clarke, S. and Foster, J. R. (2012) 'History of Blood Glucose Meters and Their Role in Self-Monitoring of Diabetes Mellitus', *British Journal of Biomedical Science* 69(2): 83–93.

Colebrook, C. (2008) 'On Not Becoming Man: The Materialist Politics of Unactualized Potential', in Alaimo, S. and Heckman, S. (eds), *Material Feminisms*. Indiana University Press.

Cote, M., Gerbaudo, P. and Pybus, J. (2016) 'Politics of Big Data', *Digital Culture and Society* 2(2): 12–24.

Cook-Deegan, R. (1991) 'The Origins of the Human Genome', *The FASEB Journal* 5: 9–11.

Cook-Deegan, R. (1994) *The Gene Wars: Science, Politics and the Human Genome*. W. W. Norton & Co.

Coole, D. and Frost, S. (2010) *New Materialisms: Ontology, Agency, and Politics*. Duke University Press.

Cooper, M. (2008) *Life As Surplus: Biotechnology and Capitalism in the Neoliberal Era*. University of Washington Press.

Couldry, N. and Hepp, A. (2013) 'Conceptualising Mediatization: Contexts, Traditions, Arguments', *Communication Theory* 23(3): 191–202.

Crary, J. (2000) *Suspensions of Perception: Attention, Spectacle and Modern Culture*. MIT Press.

Crawford, K., Lingel, J. and Karppi, T. (2015) 'Our Metrics, Ourselves: A Hundred Years of Self-tracking from the Weight Scale to the Wrist Wearable Device', *European Journal of Cultural Studies* 18(4): 479–96.

Crick, F. (1958) 'On Protein Synthesis', in Sanders, F. K. (ed.), *Symposia of the Society for Experimental Biology, Number XII: The Biological Replication of Macromolecules*. Cambridge University Press.

Cubonicks, L. (2015) XF Xenofeminism: A Politics for Alienation, www. laboriacuboniks.net @alienation.

Data Team (2015) 'Syria's Drained Population', *The Economist*, 30 September.

Dean, J. (2005) 'Communicative Capitalism: Circulation and the Foreclosure of Politics', *Cultural Politics* 1(1): 51–74.

Dean, J., Anderson, J. and Lovink, G. (eds) (2013) *Reformatting Politics: Information Technology and Global Civil Society*. Routledge.

De Beauvoir, S. (1949) *The Second Sex*. Everyman.

DECC (2012) *Smart Metering Implementation Programme: First Annual Progress Report*. Department of Energy and Climate Change, Crown Copyright.

DECC (2014) *Smart Grid Vision and Routemap*. Department of Energy and Climate Change, Crown Copyright.

De Costa, B. (2008) *Tactical Biopolitics: Art Activism and Technoscience*. MIT Press.

Delphy, C. (1976) 'Pour un feminisme materialistes', *L'Arc* 61: 197–8.

Deleuze, G. (1992) 'Postscript on the Societies of Control', *October* 59 (Winter): 3–7.

Deleuze, G. and Guattari, F. (2001) *A Thousand Plateaus*. Bloomsbury.

Department of Health, NHS England et al. (2014) 'Human Genome: UK to Become World Number 1 in DNA Testing', Press Release, 1 August, www. gov.uk/government/news/human-genome-uk-to-become-world-number-1-in-dna-testing.

Deuze, M. (2012) *Media Life*. Polity Press.

Devine-Wright, P., Devine-Wright, H. and Sherry-Brennan, F. (2010) 'Visible Technologies, Invisible Organisations: An Empirical Study of Public Beliefs About Electricity Supply Networks', *Energy Policy* 38: 4127–34.

Dick, P. K. (1962) *The Man in the High Castle*. Putnam.

Doyle, J. (2007) 'Picturing the Clima(c)tic: Greenpeace and the Representational Politics of Climate Change Communication', *Science as Culture* 16(2): 129–50.

Dunne, A. and Raby, F. (2013) *Speculative Everything*. MIT Press.

Duster, T. (1990) *Eugenics by the Backdoor*. Routledge.

Dutton, W. (1999) *Society on the Line: Information Politics in the Digital Age*. Oxford University Press.

Dyer-Witheford, N. (1999) *Cyber Marx: Cycles and Circuits of Struggle in High-Technology Capitalism*. University of Illinois Press.

Dyer-Witheford, N. (2015) *Cyber-proletariat: Life in the Digital Vortex*. Pluto.

Edwards, P. (2010) *A Vast Machine: Computer Models, Climate Data, and the Politics of Global Warming*. MIT Press.

Ehn, P., Nisson, E. M. and Topgaard, R. (2014) *Making Futures: Marginal Notes on Innovation, Design, and Democracy*. MIT Press.

Endersby, J. (2007) *A Guinea Pig's History of Biology*. Arrow Books.

Ezzell, C. (2000) 'The Business of the Human Genome', *Scientific American* 283(1): 49.

Featherstone, D. (2013) 'The Contested Politics of Climate Change and the Crisis of Neo-liberalism', *ACME: An International E-Journal for Critical Geographies* 12(1): 44–64.

Ferris, E., Cernea, M. and Petz, D. (2011) *On the Front Line of Climate Change and Displacement: Learning from and with Pacific Island Countries*. The Brookings Institute, LSE, London.

Fitbit (2014) Privacy Policy, www.fitbit.com.

Fortun, M. (2008) *Promising Genomics: Iceland and deCODE Genetics in a World of Speculation*. University of California Press.

Foster, I. et al. (2008) 'Cloud Computing and Grid Computing 360-Degree Compared', Ioan Raicu Distributed Systems Laboratory Computer Science Department, University of Chicago, https://arxiv.org/ftp/arxiv/papers/0901/0901.0131.pdf.

Fotopoulou, A. and O'Riordan, K. (2016) 'Training to Self-Care: Fitness Tracking, Biopedagogy and the Healthy Consumer', *Health Sociology Review* 25(3): 54–68.

Franklin S. (1997) *Embodied Progress: A Cultural Account of Assisted Conception*. Routledge.

Franklin, S. (2000) 'Global Nature and the Genetic Imaginary', in Franklin, S., Lury, C. and Stacey, J., *Global Nature, Global Culture*. Sage.

Franklin, S. (2006) 'The Cyborg Embryo: Our Path to Transbiology', *Theory, Culture and Society* 23(7–8): 167–87.

Franklin, S. (2007) *Dolly Mixtures: The Remaking of Genealogy*. Duke University Press.

Franklin, S. (2013) *Biological Relatives: IVF, Stem Cells and the Future of Kinship*. Duke University Press.

Fraser, M., Kember, S. and Lury, C. (2006) *Inventive Life: Approaches to the New Vitalism*. Sage.

Freidberg, A. (2006) *The Virtual Window: From Alberti to Microsoft*. MIT Press.

Frizzo-Barker, J. and Chow White, P. (2012) '"There's an App for That": Mediating Mobile Moms and Connected Careerists Through Smartphones and Networked Individualism', *Feminist Media Studies* 12(4): 580–9.

Fuchs, C. (2011) *Foundations of Critical Theory and Information Studies*. Routledge.

Gabrys, J. (2014) 'Powering the Digital: From Energy Ecologies to Electronic Environmentalism', in Maxwell, R. Raundalen, J. and Lager, N. V. (eds), *Media and the Ecological Crisis*. Routledge.

Gabrys, J. (2016) 'Re-thingifying the Internet of Things', in Starosielski, N. and Walker, J. (eds), *Sustainable Media: Critical Approaches to Media and Environment*. Routledge.

Galloway, A. (2004) *Protocol: How Power Exists After Decentralisation*. MIT Press.

Galloway, A. (2012) *The Interface Effect*. Polity.

Galloway, A. (2014) 'The Cybernetic Hypothesis', *Differences* 25(1): 107–31.

Galloway, A., Thacker, E. and Wark, M. (2014) *Excommunication: Three Inquiries into Media and Mediation*. University of Chicago Press.

Garcia-Sancho, M. (2012) *Biology, Computing and the History of Molecular Sequencing*. Palgrave Macmillan.

Gartenberg, D. et al. (2013) 'Collecting Health-related Data on the Smart Phone: Mental Models, Cost of Collection, and Perceived Benefit of Feedback', *Personal and Ubiquitous Computing* 17(3): 561–70.

Gehl, R. W. (2014) *Reverse Engineering Social Media: Software, Culture and Political Economy in New Media Capitalism*. Temple University Press.

General Electric (2009) *Smart Grids at Work*, www.ge.com/sites/default/files/ge_luke_clemente_BofA_121009_0.pdf.

Gibson, M. (2002) 'The Powers of the Pokémon: Histories of Television, Histories of the Concept of Power', *Media International Australia* 104: 107–15.

Giddings, S. (2016) 'Pokemon Go as Distributed Imagination', *Mobile Media and Communications* (November), online first (n.p.).

Gillespie, T., Boczkowski, P. and Foot, K. (eds) (2014) *Media Technologies: Essays on Communication, Materiality, and Society*. MIT Press.

Gitleman, L. (2013) *Raw Data is an Oxymoron*. MIT Press.

Gleick, P. H. (2014) 'Water, Drought, Climate Change, and Conflict in Syria', *Weather, Climate, and Society* 6: 331–40.

Gregg, M. (2015) 'Inside the Data Spectacle', *Television and New Media* 16(1): 37–51.

Griziotti, G., Lovaglio, D. and Terranova, T. (2012) 'Netwar 2.0: The Convergence of Streets and Networks', *Open Democracy Net* (23 February).

Grosz, E. (2011) *Becoming Undone: Darwinian Reflections on Life, Politics, and Art*. Duke University Press.

Grusin, R. (ed.) (2015) *The Non-Human Turn*. University of Minnesota Press.

Hansen, M. (2006) 'Media Theory', *Theory, Culture & Society* 23(2–3): 297–306.

Hall, S. (1973) 'Encoding and Decoding in the Television Discourse', in Hall, S., Hobson, D., Lowe, A. and Willis, P. (eds), *Culture, Media, Language: Working Papers in Cultural Studies, 1972–79*. Centre for Contemporary Cultural Studies, Birmingham. Routledge, 1980.

Haran, J. (2007) 'Managing the Boundaries Between Maverick Cloners and Mainstream Scientists: The Life Cycle of a News Event in a Contested Field', *New Genetics and Society* 26(2): 203–19.

Haran, J. (2011a) 'The UK Hybrid Embryo Controversy: Delegitimising Counterpublics', *Science as Culture* 22(4): 567–88.

Haran, J. (2011b) 'Campaigns and Coalitions: Governance by Media', in Rödder, S., Franzen, M. and Weingart, P. (eds), *The Sciences' Media Connection – Communication to the Public and its Repercussions*. Sociology of the Sciences Yearbook, Dordrecht. Springer.

Haran, J. and Kitzinger, J. (2010) 'Modest Witnessing and Managing the Boundaries Between Science and the Media: A Case Study of Breakthrough and Scandal', *Public Understanding of Science* 18(6): 634–52.

Haran, J., Kitzinger, J., McNeil, M. and O'Riordan, K. (2007) *Human Cloning and the Media*. Routledge.

Haraway, D. (1985) 'A Manifesto for Cyborgs: Science, Technology, and Socialist Feminism in the 1980s', *Socialist Review* 15(2): 65–107.

Haraway, D. (1988) 'Situated Knowledges: The Science Question in Feminism and the Privilege of Partial Perspective', *Feminist Studies* 14(3): 575–99.

Haraway, D. (1992) 'The Promises of Monsters: A Regenerative Politics for Inappropriate/d Others', in Grossberg, L., Nelson, C. and Treichler, P. (eds), *Cultural Studies*. Routledge.

Haraway, D. (1997) *Modest–Witness@Second–Millennium. FemaleMan–Meets–OncoMouse: Feminism and Technoscience*. Routledge.

Haraway, D. (2007) *When Species Meet*. University of Minnesota Press.

Hayles, N. K. (1999) *How We Became Posthuman: Virtual Bodies in Cybernetics, Literature, and Informatics*. University of Chicago Press.

Hedgecoe, A. (2004) *The Politics of Personalised Medicine: Pharmacogenetics in the Clinic*. Cambridge University Press.

Hennessy, R. (1993) *Materialist Feminism and the Politics of Discourse*. Routledge.

Hennessey, R. and Ingraham, C. (1997) *Materialist Feminism: A Reader in Class, Difference, and Women's Lives*. Routledge.

Herper, M. (2014) 'Flatley's Law: The Company Speeding A Genetic Revolution', *Forbes*, 20 August.

Higuchi, T. (2010) 'Atmospheric Nuclear Weapons Testing and the Debate on Risk Knowledge in Cold War America, 1945–1963', in McNeill, J. R. (ed.), *Environmental Histories of the Cold War*. Cambridge University Press.

Hinton, P. and Van der Tuin, I. (2014) 'Preface', *Women: A Cultural Review*, 25: 1–8.

Hopkins, M., Martin, P., Nightingale, P., Kraft, A. and Mahdi, S. (2007) 'The Myth of the Biotech Revolution: An Assessment of Technical, Clinical and Organizational Change', *Research Policy* 36(4): 566–89.

House of Lords Select Committee on Science and Technology (2000). Science and Society: Third Report of the Session 1999–2000. London: HMSO.

House of Lords (2013) Lords Hansard, Daily Hansard, 28 February 2013: Column 1155, www.publications.parliament.uk.

Hui, Y. (2015) 'Towards A Relational Materialism: A Reflection on Language, Relations and the Digital', *Information: Digital Culture & Society* 1(1): 131–48.

IEA (International Energy Agency) (2011) *Technology Roadmap: Smart Grids*, www.iea.org.

Ingold, T. (2012) 'Towards an Ecology of Materials', *Annual Review of Anthropology* 41: 427–42.

Ironstone, P. (2011) 'Narrating the Coming Pandemic: Pandemic Influenza, Anticipatory Anxiety, and Neurotic Citizenship', in Crosthwaite, P. (ed.), *Criticism, Crisis, and Contemporary Narrative: Textual Horizons in an Age of Global Risk*. Routledge.

Jackson, S. (2001) 'Why a Materialist Feminism is (Still) Possible – and Necessary', *Women's Studies International Forum* 24(3/4): 283–93.

Jarrett, K. (2015) *Feminism, Media and Digital Labour: The Digital Housewife*. Routledge.

Jasanoff, S. (2005) *Designs on Nature: Science and Democracy in Europe and the United States*. Princeton University Press.

Jensen, L. (2009) *The Rapture*. Bloomsbury.

Johnson, S. (1999) *Interface Culture: How New Technology Transforms the Way We Create and Communicate*. Basic Books.

Jones, G. (2006) *Life*. Aqueduct Press.

Jordan, T. (1999) *Cyberpower: The Culture and Politics of Cyberspace*. Routledge.

Jordan, T. (2004) 'The Pleasures and Pains of Pikachu', *European Journal of Cultural Studies* 7(4): 461–80.

Jordan, T. (2013a) 'Information as Politics', *Culture Machine* 14: 1–22.

Jordan, T. (2013b) *Internet, Society and Culture: Communicative Practices Before and After the Internet*. Bloomsbury.

Jordan, T. (2015) *Information Politics: Liberation and Exploitation in the Digital Society*. Pluto.

Jordanova, L. (1989) *Sexual Visions: Images of Gender in Science and Medicine Between the Eighteenth and Twentieth Centuries*. University of Wisconsin Press.

Kay, L. (2000) *Who Wrote the Book of Life? A History of the Genetic Code*. Stanford University Press.

Keller, E. (2002) *The Century of the Gene*. Harvard University Press.

Kelty, C. (2008) *Two Bits: The Cultural Significance of Free Software*. Duke University Press.

Kember, S. (2015) *iMedia: The Gendering of Objects, Environments and Smart Materials*. Palgrave.

Kember, S. and Zylinksa, J. (2012) *Life After New Media*. MIT Press.

Kennedy, H. (2015) 'Standards, Values and (Better Thinking About) Power in Creative Digital Work', ECREA Keynote DCC Workshop: Digital Culture: Standards, Disruptions and Values, University of Salzburg, 26–7 November.

Kennedy, H., Poell, T. and Van Dijck, J. (2015) 'Introduction: Data and Agency', *Big Data and Society* 2(2): 1–7.

Keogh, B. (2016) 'Pokémon Go, the Novelty of Nostalgia, and the Ubiquity of the Smartphone', *Mobile Media and Communication* (November), online first (n.p.).

Kerr, A. and Shakespeare, T. (2002) *Genetic Politics from Genetics to Genome*. Clarion.

Kim, D. and Kim, H. (2014) 'Biosensor Interface: Interactive Media Art Using Biometric Data', *International Journal of Bio-Science and Bio-Technology* 6(1): 129–36, www.sersc.org/journals/IJBSBT/vol6_no1/14.pdf.

Kirby, D. (2011) *Lab Coats in Hollywood: Science, Scientists, and Cinema*. MIT Press.

Kirby, V. (1997) *Telling Flesh the Substance of the Corporeal*. Routledge.

Kirby, V. (2011) *Quantum Anthropologies: Life at Large*. Duke University Press.

Kitchin, R. (2014) *The Data Revolution: Big Data, Open Data, Data Infrastructures and Their Consequences*. Sage.

Kittler, F. (1986) *Grammophon Film Typewriter*. Brinkmann & Bose.

Kitzinger, J. (2006) 'Constructing and Deconstructing the "Gay Gene": Media Reporting of Genetics, Sexual Diversity and "Deviance"', in Ellison, G. T. H. and Goodman, A. H. (eds), *The Nature of Difference: Science, Society and Human Biology*. Taylor and Francis.

Krause, B. (2015) *Voices of the Wild: Animal Songs, Human Din, and the Call to Save Natural Soundscapes*. Yale University Press.

Latour, B. (1991) 'Technology is Society Made Durable in Law', in *A Sociology of Monsters: Essays on Power, Technology and Domination*. Routledge.

Latour, B. (1996) *Aramis or the Love of Technology*. Harvard University Press.

Latour, B. and Weibel, P. (eds) (2005) *Making Things Public – Atmospheres of Democracy*. MIT Press.

Laurel, B. (1990) *The Art of Human/Computer Interface Design*. Addison Wesley.

Law, J. and Mol, A. (1995), 'Notes on Materiality and Sociality', *The Sociological Review* 43: 274–94.

Lemke, T. (2016) 'Rethinking Biopolitics: The New Materialism and the Political Economy of Life', in Wilmer, S. and Zukauskaite, A. (eds), *Resisting Biopolitics: Philosophical, Political, and Performative Strategies*, Routledge.

Ley, W. (1964) *Engineers' Dreams*. Viking (revised from 1954).

Lievrouw, L. (2014) 'Materiality and Media in Communication Technology Studies', in Gillespie, T., Boczkowski, P. and Foot, K. (eds), *Media Technologies: Essays on Communication, Materiality, and Society*. MIT Press.

Lilliestam, J. and Ellenbeck, S. (2011) 'Energy Security and Renewable Electricity Trade: Will Desertec Make Europe Vulnerable to the "Energy Weapon"?', *Energy Policy* 39(6): 3380–91.

Lindee, S. and Nelkin, D. (1995) *The DNA Mystique: The Gene as a Cultural Icon*. W.H. Freeman.

Littlefield, M. (2008) 'Constructing the Organ of Deceit: The Rhetoric of fMRI and Brain Fingerprinting in Post-9/11 America', *Science, Technology and Human Values* 34(3): 365–92.

Livingstone, S. (2009) 'Enabling Media Literacy for "Digital Natives": A Contradiction in Terms?', in *'Digital Natives': A Myth?*, POLIS, London School of Economics and Political Science.

Love, H. (2010) 'Close But Not Deep: Literary Ethics and the Descriptive Turn', *New Literary History* 41: 371–91.

Lupton, D. (2013) 'Digitized Health Promotion: Personal Responsibility for Health in the Web 2.0 Era', Sydney Health & Society Group Working Paper No. 5.

Lupton, D. (2016) *The Quantified Self*. Polity Press

Lyon, D. (1988) *The Information Society: Issues and Illusions*. Polity Press.

Lyon, D. (1994) *The Electronic Eye: The Rise of the Surveillance Society*. University of Minnesota Press.

Lyon, D. (2001) *Surveillance Society: Monitoring Everyday Life*. Open University Press.

Lyon, D. (2006) *Theorizing Surveillance: The Panopticon and Beyond*. Routledge.

Mackenzie, A. (2002) *Transductions: Bodies and Machines at Speed*. Continuum.

Mackenzie, A. (2010) 'Bringing Sequences to Life: How Bioinformatics Corporealizes Sequence Data', *New Genetics and Society* 22(3): 315–32.

Mackenzie, A. (2010) *Wirelessness: Radical Empiricism in Network Cultures*. MIT Press.

Mackenzie, A. (2013) 'From Validating to Verifying: Public Appeals in Synthetic Biology'. *Science as Culture*. 22(4): 476–96.

McLuhan, M. (1964) *Understanding Media: The Extensions of Man*. McGraw-Hill.

McNally, R. and Glasner, P. (2007) 'Survival of the Gene? Twenty-first-century Visions from Genomics, Proteomics and the New Biology', in Glasner, P., Atkinson, P. and Greeslade, H. (eds), *New Genetics, New Social Formations*. Routledge.

McNeil, M. and Haran, J. (2013) 'Publics of Bioscience', *Science as Culture* 22(4): 433–51.

McNeil, M. (2007) *Feminist Cultural Studies of Science and Technology*. Routledge.

McNeil, M. (2013) 'Between a Rock and a Hard Place: The Deficit Model, the Diffusion Model and Publics in STS', *Science as Culture* 22(4): 589–608.

Magnet, S. (2011) *When Biometrics Fail*. Duke University Press.

Mahdawi, A. (2013) 'Your Body Isn't a Temple, It's a Data Factory Emitting Digital Exhaust', *Guardian*, 25 January.

Mamo, L. (2007) *Queering Reproduction: Achieving Pregnancy in the Age of Technoscience*. Duke University Press.

Mann, S., Nolan, J. and Wellman, B. (2003) 'Sousveillance: Inventing and Using Wearable Computing Devices for Data Collection in Surveillance Environments', *Surveillance & Society* 1(3): 331–55.

Martin, P., Hopkins, M., Nightingale, P. and Craft, A. (2009) 'On a Critical Path: Genomics, the Crisis of Pharmaceutical Productivity and the Search for Sustainability', in Atkinson, P., Glasner, P. and Locke, M. (eds), *The Handbook of Genetics and Society: Mapping the New Genomic Era*. Routledge.

Massanari, A. (2015) *Participatory Culture, Community, and Play: Learning from Reddit*. Peter Lang.

Matter, J. et al. (2016) 'Rapid Carbon Mineralization for Permanent Disposal of Anthropogenic Carbon Dioxide Emissions', *Science* 352(6291): 1312–14.

M'Charek, A. (2005) *The Human Genome Diversity Project: An Ethnography of Scientific Practice*. Cambridge University Press.

Mellor F. (2009) 'The Politics of Accuracy in Judging Global Warming Films', *Environmental Communication: A Journal of Nature And Culture* 3: 134–50.

Middleton, A., Wright, C., Morley, K., Bragin, E., Firth, H. and Parker, M. (2015) 'Potential Research Participants Support the Return of Raw Sequence Data', *Journal of Medical Genetics* 52(8): 571–4.

Middleton, A. (2016) 'Socialising the Genome', 13 March 2016, www.genomicsengland.co.uk/socialising-the-genome-blog.

Milne, E. (2013) *Letters, Postcards, Email: Technologies of Presence*. Routledge.

Mitchell, R. and Waldby, C. (2006) *Tissue Economies: Blood, Organs, and Cell Lines in Late Capitalism*. Duke University Press.

Mol, A. (2012) 'Mind Your Plate! The Ontonorms of Dutch Dieting', *Social Studies of Science* 43(3): 379–96.

Moretti, F. (2000) 'Conjectures on World Literature', *New Left Review* 1, Jan–Feb: 54–68.

Morrison, T. (1987) *Beloved: A Novel*. Alfred Knopf.

Mort, M., Finch, T. and May, K. (2009) 'Making and Unmaking Telepatients: Identity and Governance in New Health Technologies', *Science, Technology and Human Values* 34(1): 9–33.

Morton, T. (2013) *Hyperobjects: Philosophy and Ecology After the End of the World*. University of Minnesota Press.

Mosco, V. (2014) *To the Cloud: Big Data in a Turbulent World*. Paradigm.

Mullaney, T. (2015) 'Opinion: Why I Hate the Fitbit IPO (and You Should, Too)', *MarketWatch*, 18 June.

Munro, A. (2015) 'Orkney Wind Turbine Spins into Record Books', *The Scotsman*, 15 April.

Munster, A. (2011) *Materialising New Media: Embodiment in Information Aesthetics*. University Press of New England.

National Infrastructure Commission (2016) *Smart Power*, www.gov.uk/government/publications/smart-power-a-national-infrastructure-commission-report

Neate, R. (2016) 'Fitbit Stock Sinks After Company Warns Shareholders Over Profits', *Guardian*, 23 February.

Nerlich, B. and Hellsten, I. (2004) 'Genomics: Shifts in Metaphorical Landscape Between 2000 and 2003', *New Genetics and Society* 23(3): 255–68.

Oliver, J., Savičić, G. and Vasiliev, D. (2011) *The Critical Engineering Manifesto*. The Critical Engineering Working Group Berlin, https://criticalengineering.org.

Olson, P. and Tilley, A. (2014) 'The Quantified Other: NEST and Fitbit Chase, a Lucrative Side Business', *Forbes*, 17 April.

O'Riordan, K. (2008) 'Human Cloning in Film: Horror, Ambivalence, Hope', *Science as Culture* 17(2): 145–62.

O'Riordan, K. (2010) *The Genome Incorporated: Constructing Biodigital Identity*. Ashgate.

O'Riordan, K. (2011) 'Writing Biodigital Life: Personal Genomes and Digital Media', *Biography* 34(1): 119–31.

O'Riordan, K. (2013) 'Biodigital Publics: Personal Genomes as Digital Media Artefacts', *Science as Culture* 22(4): 516–39.

O'Riordan, K., Fotopoulou, A. and Stephens, N. (2016) 'The First Bite: Imaginaries, Promotional Publics and the Laboratory Grown Burger', *Public Understanding of Science*, 29 March.

O'Riordan, K., Parker, J., Devereaux, E. and Harris, D. (2017) 'Making Sense of Sensors', *Digital Culture and Society*, Special Issue on Hacking and Making (forthcoming).

Packer, J. and Crofts Wiley, S. (2014) 'Strategies for Materializing Communication', in Gillespie, T., Boczkowski, P. and Foot, K. (eds), *Media Technologies: Essays on Communication, Materiality, and Society*. MIT Press.

Papacharissi, Z. (2015) *Affective Publics: Sentiment, Technology, Politics*. Oxford University Press.

Parisi, L. (2004) *Abstract Sex: Philosophy, Biotechnology and the Mutations of Desire*. Continuum.

Parmentier, F-J. W. et al. (2015) 'Rising Methane Emissions from Northern Wetlands Associated with Sea Ice Decline', *Geophysical Research Letters* 42(17): 7214–22.

Parikka, J. (2015) *A Geology of New Media*. University of Minnesota Press.

Parry, B. (2004) *Trading in the Genome: Investigating the Commodification of Bio-information*. Columbia University Press.

Pink, S. and Ardevol, E. (2016) *Digital Materialities: Design and Anthropology*. Bloomsbury.

Plant, S. (1998) *Zeroes and Ones: Digital Women and the New Techoculture*. Fourth Estate.

Poster, M. (1990) *The Mode of Information: Poststructuralism and Social Context*. University of Chicago Press.

Power, N. (2015) 'Decapitalism, Left Scarcity, and the State', *Fillip* 20, https://fillip.ca/content/decapitalism-left-scarcity-and-the-state.

Power, N. (2016) 'Philosophy, Sexism, Emotion, Rationalism', in Kolozova, K. and Joy, E. (eds), *After the 'Speculative Turn': Realism, Philosophy, and Feminism*. Punctum Books.

Preciado, P. (2013) *Testo Junkie: Sex, Drugs, and Biopolitics in the Pharmaco-pornographic Era*. Feminist Press.

Preda, A. (1999) 'The Turn to Things: Arguments for a Sociological Theory of Things', *Sociological Quarterly* 40(2): 347–66.

Rabinow, P. (1999) *French DNA: Trouble in Purgatory*. University of Chicago Press.

Reardon, J. (2005) *Race to the Finish: Identity and Governance in an Age of Genomics*. Princeton University Press.

Reardon, J. (2012) 'The Democratic, Anti-Racist Genome? Technoscience at the Limits of Liberalism', *Science as Culture* 21(1): 25–47.

Reardon, J. (2014a) 'The Post-Genomic Condition: Ethics, Justice, Knowledge After the Genome', SBS Seminar Series, UCSF, 10 February.

Reardon, J. (2014b) 'Genomic Cosmopolitanism and the Re-Constitution of the Nation-State: The Case of Generation Scotland', Science, Technology and Innovation Series, University of Edinburgh, 29 September.

Reardon, J. (2017) *The Postgenomic Condition: Ethics, Justice, Knowledge After the Genome*. University of Chicago Press.

Rennie, J. (2000) 'Bracing for the Imminent', Editorial, *Scientific American* 283(1): 6.

Reuters (2014) 'Desertec Shareholders Jump Ship as Solar Project Folds', 14 October, www.reuters.com/article/germany-desertec-idUSL6N0S535V20141014.

Robins, K. and Webster, F. (1988) 'Cybernetic Capitalism: Information, Technology, Everyday Life', in Mosko, V. and Wasko, J. (eds), *The Political Economy of Information*. University of Wisconsin Press.

Roof, J. (2007) *The Poetics of DNA*. University of Minnesota Press.

Rose, H. (2001) *The Commodification of Bioinformation: The Icelandic Health Sector Database*. Wellcome Trust.

Rose, H. and Rose, S. (2014) *Genes, Cells and Brains: The Promethean Promises of the New Biology*. Verso.

Rossiter, N. (2016) *Software, Infrastructure, Labor: A Media Theory of Logistical Nightmares*. Routledge.

Rothe, D. (2014) 'Energy for the Masses? Exploring the Political Logics Behind the Desertec Vision', *Journal of International Relations and Development* (September), online first (n.p.).

Salen Tekinbaş, K. (2016) 'Afraid to Roam: The Unlevel Playing Field of *Pokémon Go*', *Mobile Media and Communication* (December), online first (n.p.).

Samus, T., Lang, B. and Rohn, H. (2013) 'Assessing the Natural Resource Use and the Resource Efficiency Potential of the Desertec Concept', *Solar Energy* 87: 176–83.

Schwartz, H. (1996) *The Culture of the Copy: Striking Likenesses, Unreasonable Facsimiles*. MIT Press.

Serres, M. (1982) *The Parasite*. University of Minnesota Press.

Shakespeare, T. (1998) 'Choices and Rights: Eugenics, Genetics and Disability Equality', *Disability & Society* 13: 665–81.

Shannon, C. and Weaver, W. (1949) *The Mathematical Theory of Communication*. University of Illinois Press.

Shapiro, B. (2015) *How to Clone a Mammoth: The Science of De-Extinction*. Princeton University Press.

Sheldon, R. (2015) 'Form/Matter/Chora: Object-Oriented Ontology and Feminist New Materialism', in Grusin, R. (ed.), *The Non-Human Term*. University of Minnesota Press.

Shotwell, A. (2016) *Against Purity: Living Ethically in Compromised Times*. University of Minnesota Press.

Shukin, N. (2009) *Animal Capital: Rendering Life in Biopolitical Times*. University of Minnesota Press.

Silverstone, R. (2006) *On Media and Morality: The Rise of the Mediapolis*. Polity.

Smart Energy GB (2015) *Smart Energy GB Annual Report 2015*, www.smartenergygb.org/en/about-smart-energy-gb/annual-report.

Smart Energy GB (2016) 'An Ode to the Meter', press release, https://www.smartenergygb.org/en/resources/press-centre/press-releases-folder/ode-to-meter-april-2016

Smart Grid GB (2012) Smart Grid: A Race Worth Winning. A Report on the Economic Benefits of Smart Grid. Ernst and Young.

Stacey, J. (2010) *The Cinematic Life of the Gene*. Duke University Press.

Steele, R. and Clarke, A. (2013) 'The Internet of Things and Next-generation Public Health Information Systems', *Communications and Network* 5(3B1): 5–9.

Stengers, I. (2011) 'Wondering about Materialism', in Bryant, L. R., Srnicek, N. and Harman, G. (eds), *The Speculative Turn: Continental Materialism and Realism*. Re.Press.

Stephens, N. (2010) 'In Vitro Meat: Zombies on the Menu?', *SCRIPTed* 7(2): 394–401.

Stephens, N. (2013) 'Growing Meat in Laboratories: The Promise, Ontology, and Ethical Boundary-Work of Using Muscle Cells to Make Food', *Configurations* 21(2): 159–81.

Sterne, J. (2003) *The Audible Past: Cultural Origins of Sound Reproduction*. Duke University Press.

Stevens, C. and Bryan, A. (2012) 'Rebranding Exercise: There's an App For That', *American Journal of Health Promotion* 27(2): 69–70.

Stevens, H. (2013) *Life Out of Sequence: A Data-Driven History of Bioinformatics*. University of Chicago Press.

Stevens, J. (2008) 'Biotech Patronage and the Making of Homo DNA', in Da Costa, B. and Phillip, K. (ed.), *Tactical Biopolitics: Art, Activism, and Technoscience*. MIT Press.

Stiegler, B. (2010) *What Makes Life Worth Living: On Pharmacology*. Polity.

Strathern, M. (1992) *Reproducing the Future: Essays on Anthropology, Kinship and the New Reproductive Technologies*. Routledge.

Sunder Rajan, K. (2002) *Biocapital the Constitution of Postgenomic Life*. Duke University Press.

Tallbear, K. (2013) *Native American DNA: Tribal Belonging and the False Promise of Genetic Science*. University of Minnesota Press.

Terranova, T. (2004) *Network Culture: Politics for the Information Age*. Pluto.

Terranova, T. (2009) 'Another Life: The Nature of Political Economy in Foucault's Genealogy of Biopolitics', *Theory, Culture & Society* 26(6): 234–62.

Terranova, T. (2014) 'The (European) Posthuman Predicament: Rosi Braidotti's *The Posthuman* and the Future of the Humanities', *Anglistica AION* 18(2): 191–8.

Thacker, E. (2003) 'What is Biomedia?', *Configurations* 11(1): 47–79.

Thacker, E. (2004) *Biomedia*. University of Minnesota Press.

Thacker, E. (2010) 'Biomedia', in Mitchell, M. and Hanson, M. (eds), *Critical Terms for Media Studies*. University of Chicago Press.

Thomas, S. (2013) *Technobiophilia: Nature and Cyberspace*. Bloomsbury.

Thompson, C. (2005) *Making Parents: The Ontological Choreography of Reproductive Technologies*. MIT Press.

Thompson, C. (2015) 'CRISPR: Move Beyond Differences', *Nature* 522(7557): 415.

Throsby, K. (2004) *When IVF Fails: Feminism, Infertility and the Negotiation of Normality*. Palgrave Macmillan.

Tiainen, M., Katve-Kaisa, K. and Hongisto, I. (2015) 'Movement, Aesthetics, Ontology', *Cultural Studies Review* 21(2): 4–13.

Tissue Culture and Arts Project (2001) 'Pig Wings – The Chiropteran Version'.

Turner, F. (2014) 'The World Outside and the Pictures in our Networks', in Gillespie, T., Boczkowski, P. and Foot, K. (eds), *Media Technologies: Essays on Communication, Materiality, and Society*. MIT Press.

Turney, J. (2007) *Engaging Science: Thoughts, Deeds, Analysis and Action*. Wellcome Trust.

Tutton, R. (2011) 'Promising Pessimism: Reading the Futures to be Avoided in Biotech', *Social Studies of Science* 41(3): 411–29.

Tutton, R. (2016) *Genomics and the Reimagining of Personalised Medicine*. Routledge.

UNHCR (2015) *Global Trends: Forced Displacement in 2015*. United Nations High Commissioner for Refugees.

USAC of Engineers (2009) *Alaska Baseline Erosion Assessment: Study Findings and Technical Report*, http://climatechange.alaska.gov/docs/iaw_USACE_erosion_rpt.pdf.

Vasiliou, S. et al. (2016) 'CRISPR-Cas9 System: Opportunities and Concerns', *Clinical Chemistry* 62(10): 1304–11.

Van der Sluijs, J., Rommetveit, K. et al. (2015) *Policy Recommendations: Towards Socially Robust Smart Grids*. The EPINET Consortium, www.epinet.no/sites/all/themes/epinet_bootstrap/documents/smart_grid_policy_recommendations.pdf.

Van der Tuin, I. and Dolphijn, R. (eds) (2012) *New Materialism: Interviews and Cartographies*. Open Humanities Press.

Van Dijck, J. (1998) *Imagenation: Popular Images of Genetics*. Palgrave.

Van Dooren, T. (2014) *Flight Ways: Life and Loss at the Edge of Extinction*. Columbia University Press.

Van Dooren, T. and Rose, D. (2014) 'Keeping Faith with the Dead: Mourning and De-extinction', *Australian Zoologist*. Online first, unpaginated.

Venter, C. (2007) *A Life Decoded: My Genome: My Life*. Penguin.

Viola, J., Lal, B. and Grad, O. (2003) *The Emergence of Tissue Engineering as a Research Field*. Report Prepared for The National Science Foundation: Arlington, Virginia, www.nsf.gov/pubs/2004/nsf0450/start.htm.

Von Schomberg, R. and Funtowicz, S. (2007) 'Foresight Knowledge Assessment', *International Journal of Foresight and Innovation Policy* 3(1): 53–75.

Wahl-Jorgenson, K., Sambrook, R., Berry, M. et al. (2013) 'BBC Breadth of Opinion Review Content Analysis', School of Journalism, Media and Cultural Studies, Cardiff University.

Walker, P. (2012) 'DNA of 100,000 People to be Mapped for NHS', *Guardian*, 10 December, www.theguardian.com/science/2012/dec/10/1000000-peoples-dna-mapped.

Watts, L. (2008) Orkney Landscapes of Future Resistance 4S/EASST 2008. Erasmus University Rotterdam, August 2008, http://eprints.lancs.ac.uk/11452/1/watts_futureresistance.pdf.

Watts, L. (2016) 'The Electric Nemesis: Making Energy Futures Without Hubris' (Draft). Presented at Electrifying Anthropology, Durham University, http://sand14.com/wp-content/uploads/2016/03/Watts-ElectricNemesis-final.pdf.

Watts, L., Ehn, P. and Suchman, L. (2014) 'Prologue', in Ehn, P., Nilsson, E. M. and Topgaard, R. (eds), *Making Futures: Marginal Notes on Innovation, Design, and Democracy*. MIT Press.

Wellcome Trust (2014) 'Wellcome Trust Invests £27m in World-class Sequencing Facility for Genomics England and Sanger Institute', Press release, 1 August, https://wellcome.ac.uk/press-release/wellcome-trust-invests-%C2%A327m-world-class-sequencing-facility-genomics-england-and

Wellman, B. (2002) 'Little Boxes, Glocalization, and Networked Individualism', in Tanabe, M., Van Den Besselaar, P. and Ishida, T. (eds), *Digital Cities II: Computational and Sociological Approaches*. Springer.

Welsh, I. and Wynne, B. (2013) 'Science, Scientism and Imaginaries of Publics in the UK: Passive Objects, Incipient Threats', *Science as Culture* 22(4): 540–66.

Wertheim, M. (1999) *The Pearly Gates of Cyberspace: A History of Space from Dante to the Internet.* W. W. Norton & Company.

The White House, Office of the Press Secretary (2000) Remarks Made by the President, Prime Minister Tony Blair of England (via satellite), Dr. Francis Collins, Director of the National Human Genome Research Institute, and Dr. Craig Venter, President and Chief Scientific Officer, Celera Genomics Corporation, on the Completion of the First Survey of the Entire Human Genome Project. Archived at: www.genome.gov/10001356/june-2000-white-house-event.

Williams, A. and Srnicek, N. (2013) '#ACCELERATE MANIFESTO for an Accelerationist Politics', *Critical Legal Thinking: Law and the Political*, http://criticallegalthinking.com/2013/05/14/accelerate-manifesto-for-an-accelerationist-politics.

Williams, R. (1974) *Television: Technology and Cultural Form.* Routledge.

Winship, J. (1987) *Inside Women's Magazines.* Unwin Hyman.

Wolfe, C. (2009) *What is Posthumanism?* Minnesota University Press.

Wynne, B. (2006) 'Public Engagement as a Means of Restoring Public Trust in Science: Hitting the Notes, but Missing the Music?', *Community Genetics* 9(3): 211–20.

Zhuchenko, O. et al. (1997) 'Autosomal Dominant Cerebellar Ataxia (SCA6) Associated with Small Polyglutamine Expansions in the Alpha 1A-voltage-dependent Calcium Channel', *Nature Genetics* 15(1): 62–9.

Zielinski, S. (2008) *Deep Time of the Media: Toward an Archaeology of Hearing and Seeing by Technical Means.* MIT Press.

Zimmer, C. (2013) 'Bringing Them Back to Life. The Revival of an Extinct Species is No Longer a Fantasy. But is it a Good Idea?', *National Geographic* (April).

Zimmerman, E. (2014) '50 Smartest Companies: Illumina', *MIT Technology Review*, 18 February.

Zylinksa, J. (2009) *Bioethics in the Age of New Media.* MIT Press.

Zylinksa, J. (2010) 'If it Reads it Bleeds', CRASSH Centre, University of Cambridge, www.joannazylinska.net/if-it-reads-it-bleeds.

Index